MW00916387

ALGEBRA
IN
WORDS
2

*MORE Hints, Strategies
and Simple Explanations*

Gregory P. Bullock, Ph.D.

2

Bullock, Gregory P.
Algebra in words 2: more hints, strategies and simple explanations

ISBN-13: 978-1511695992

ISBN-10: 1511695994

ASIN: B00S4JF88S

BISAC: Mathematics / Study & Teaching

First Edition

The United States of America

Table of Contents

INTRODUCTION: THE SEQUEL

This book is the continuation of *ALGEBRA IN WORDS: A Guide of Hints, Strategies and Simple Explanations*.

In 2013, I decided to go back to a project I had started a few years back, but put on the back-burner. The project involved transcribing, organizing and formatting the notes I gave to my math classes. These notes were given in a unique way, and a different perspective than traditional textbooks and instructors. I call this new perspective: *in words*, as this provides the supplemental perspective of numbers, variables, and lines of equations. Since its publication, the digital edition has sold or been lent a few thousand times, around the world, albeit mostly in the US. It became available in paperback, and became available in Barnes & Noble stores throughout the United States.

I had more notes on more algebra topics to share. The success of the first *ALGEBRA IN WORDS* motivated me to make the next installment, focusing on more advanced topics such as inequalities, absolute values, more on quadratics, radicals, circles, and more, *in words*.

This is *ALGEBRA IN WORDS 2: MORE Hints, Strategies and Simple Explanations.*

This book follows the same philosophy and tone as the first *ALGEBRA IN WORDS*, which is: clear and direct communication. One of the biggest causes of students getting confused, lost, frustrated, and low grades is due to a communication breakdown between the student and the instructor and/or the textbook. Algebra is mostly taught in a very theoretical way by these sources, but when students can't interpret the theory and lines of equations properly, learning comes to a halt. This book is the communication bridge that translates what the instructor/textbook are trying to tell you into language that is more easily understood. That even means that some topics will be unapologetically be discussed in a redundant way, to help reinforce information you need to recall. Also, this digital edition contains hyperlinks for easy topic jumping, something not easily done in a textbook.

Note: Anywhere you see an asterisk (*), it is to remind you that:
*This is explained in more depth in (the first) *ALGEBRA IN WORDS: A Guide of Hints, Strategies and Simple Explanations*

THE REAL ORDER OF OPERATIONS: GEMA

Remember, when doing calculations and mathematical manipulations, *order matters*. Below is the true and complete order of operations. I recommend using the mnemonic device and acronym GEMA instead of PEMDAS[*].

1. **G**roups – Simplify groups first, if possible, from inner to outer. A group is a set of (parentheses), [brackets], {braces}, overall numerators, overall denominators, radicands, and absolute value groups. In speaking a mathematical statement, a group is often called "the quantity (then say what is in it)."

2. **E**xponents *or* roots, whichever come first, from left to right. Remember, any root can be converted into an exponent as a *rational exponent*.

3. **M**ultiplication *or* division, whichever comes first, from left to right.

4. **A**ddition *or* subtraction, whichever comes first, from left to right.

DEFINITIONS

GCF – Abbreviation for Greatest Common Factor, also known as the Greatest Common Divisor[*].

LCD – Abbreviation for Least Common Denominator, sometimes also known as the Least Common Multiple[*].

Minuend – The number from which the subtrahend is subtracted from in a subtraction problem.

Subtrahend – The number subtracted from the minuend in a subtraction problem.
The minuend minus the subtrahend equals the difference.

Origin – The center of a graph, at point (0, 0).

Domain – The set of all possible x-values corresponding to an equation, function, or graph.

Range – The set of all possible y-values corresponding to an equation, function, or graph.

Absolute Value – A group which yields only a positive value. However, terms inside the absolute value group may be negative. Also, an absolute value group may yield zero, which carries no sign. The vertical lines "| |" that surround an absolute value group are sometimes referred to as "bars."

Constraint – A condition which must be met in order to make a mathematical statement true. A constraint often tells the permissible values of x and is often in the form of an inequality statement (but can also be an equality or a representative symbol). A mathematical (or logic) statement may have more than one constraint and may involve and/or or an "if, then" statement. Constraints are often applied in inequalities such as Systems of Linear Inequalities, Quadratic Inequalities and Rational Inequalities, which are all forms of Compound Inequalities. For more, see: Constraints with "And" & "Or".

Parent Graph – The simplest graph in a category of graph or function types. Each parent graph has an associated *Parent Function*. The parent can be thought of as a reference from which more complicated forms (transformations) of the function are made, such as translations, reflections, shrinking, stretching, etc. In the parent function, the leading term(s) that define the type of function usually have a coefficient of positive 1, with no other terms.

Root (also known as the **Index**) – Taking a *root* of a number or term is like saying "what number (or term), which, when multiplied times itself the root-number-of-times, will equal the radicand?" For instance,
$\sqrt[5]{-32} = -2$ because $(-2)^5 = -32$
In Words: The fifth root of negative thirty-two equals negative two because negative two to the fifth power equals negative thirty-two. This example can also be looked at as:
$\sqrt[5]{(-2)^5}$ when the radicand is in prime-factored exponent form. This is explained more in: The Definition of Higher Order Roots.

Radicand – The content found inside/under the radical "$\sqrt{}$" symbol. The radicand of the Quadratic Formula is known as the *discriminant*.

Discriminant – The radicand of the Quadratic Formula, which, once simplified, reveals if the solution to a quadratic equation will be either:
- one real solution (if the radicand is zero)
- two real solutions (if the radicand is a positive number)
- two complex numbers (unreal solutions, if the radicand is a negative number)

Using the discriminant is a shortcut way to determine the type of solution(s) to a quadratic equation, however, this can also be realized by solving the quadratic equation normally.

Complex Number – The combination of a real and an imaginary number, in the binomial-like form:
+/- #i
Complex numbers sometimes occur when quadratic equations are solved (by Completing the Square or the Quadratic Formula); the "sometimes" is when there is a negative number in the square-root radical. Anytime complex numbers are the result of a quadratic equation, they can and will only appear in conjugate pairs. Likewise, *the product of a pair of conjugate complex numbers results in the sum of two squares*. This is also why a complex number must be multiplied by its conjugate when rationalizing a denominator with a complex number.

Anytime a quadratic equation results in a complex number, it signifies that there are no real-number x-intercepts, and thus the resulting parabola lies above or below the x-axis.

Exact Form – A number, given in the form of certain reduced fractions, radicals, logs, or constants (like π or e), which have many decimal places, often irrational numbers or numbers with repeating decimals. These numbers are reported in "exact form" to avoid reporting them in *approximate form* with rounded decimal places.

The Quadratic Formula:

$$x = \frac{-b \pm \sqrt{b^2 - 4ac}}{2a}$$

In Words: x equals negative b, plus or minus the square root of the quantity (b-squared minus four-a-c), all-over two-a. The values for a, b and c come from the Standard Form of a Quadratic Equation: $y = ax^2 + bx + c = 0$ (where "a" is a non-zero number)

For more, see: Where Did the Quadratic Formula Come From?

The Definition of Higher Order Roots:

$$\sqrt[n]{x^n} = (\sqrt[n]{x})^n = x \qquad \text{where "n" represents the same number.}$$

In Words: When the root and power of a base are the same, the solution equals just that base. Likewise, when a root is taken to a power, where the root and power are the same, the answer is the original radicand. This is discussed in more detail in: Converting Radicals & Rational Exponents.

The Definition of a Logarithm is:
$y = \log_b x$
which is based off of, and is a rearrangement of,
The Definition of an Exponential Function:
$x = b^y$
where x is some real quantity (≥ 0) and y is the (usually unknown) variable in the exponent to be solved for. This is continued in Logarithms & Natural Logs.

SYMBOL → MEANING *IN WORDS*

∝ The "variation symbol" meaning "varies with respect to," also known as the "proportionality symbol," meaning "proportional to." This is explained more in: VARIATION & PROPORTIONALITY.

k A general symbol used as a "variation constant" (a.k.a. the proportionality constant, a.k.a. the coefficient of proportionality). This is a constant (a number with a fixed value based on the proportionality of two variables or quantities, which functions to balance out the values of those quantities when they are not equal). This is explained more in: VARIATION & PROPORTIONALITY. Note: k is used in various contexts. Be sure not to confuse this k with the k representing vertical shifts, used in standard form equations for a parabola, a circle, or any other graph. Also, there may be constants other than k in different equations.

e Euler's Number (pronounced "Oiler's Number"), often referred to as "the number e". This number is a *constant* and an irrational number well-known for being used in calculating compound interest. It is also known and used as the *Inverse Natural Log*. The first four significant figures (rounded to the thousandths place) are: 2.718

ℝ All Real Numbers. This is what the symbol may look like in books. However, when you write it yourself, draw it like a capital letter R with a second vertical line, as "|R". This is a fancy way of saying:
- "every number is a solution,"
- "any number used makes the statement true," and
- "the graph exists at all points from negative infinity to positive infinity."

Ø Null-Set, a.k.a. Empty Set. The symbol looks like a circle with a slash through it. This is a fancy way of saying:
- "no solutions," and
- "the graph isn't there."

∞ Infinity. Horizontally, this is the furthest to the right you can go on a graph.
Vertically, this is the highest up you can go on a graph. This is used in Interval Notation with the appropriate (noninclusive) parenthesis because you can never truly reach infinity.

-∞ Negative Infinity. Horizontally, this is the furthest to the left you can go on a graph.
Vertically, this is the lowest down you can go on a graph. This is used in Interval Notation with the appropriate (noninclusive) parenthesis because you can never truly reach negative infinity.

≥ Greater-than-or-equal-to; an *inclusive* symbol, also meaning:
   At least
   A Minimum of
   No less than
   Down-to-and-including
   Without going under

≤ Less-than-or-equal-to; an *inclusive* symbol, also meaning:
   At most
   A Maximum of
   No more than
   Up-to-and-including
   Without going over

> Greater-than (only); a *noninclusive* symbol, also meaning:
   Any amount greater than
   Staying over
   Can be any amount more than
   Can't be (the said number) or less

< Less-than (only); a *noninclusive* symbol, also meaning:
   Any amount under
   Staying under
   Can be any amount less than
   Can't be (the said number) or more

The larger, or open, side of the inequality symbol represents the greater side, however, when written or when reading, you read the symbol as it is presented, from left to right. For instance, you would read "4 > x" as "Four is greater an x." Usually the variable is written on the left side, so this could be flipped to "x < 4" which would be more properly read as "x is less than 4."

For more on setting up inequality symbols in relation to zero, positives and negatives, see: Constraints with "And" & "Or".

INCLUSIVE vs. NONINCLUSIVE

Inclusive means: Can include.
- An inequality with the "or-equal-to" line under it is "inclusive."
- A **Closed-dot** (a.k.a solid, filled-in dot) or **bracket**: [,] means: touches this point.
- For Linear Inequalities, a solid line is used to mean "inclusive."

Noninclusive means: Does not include.
- An inequality *without* the "or-equal-to" line under it is "noninclusive."
- A **Open-dot** or **parenthesis** (,) means: approaches, and gets infinitely close, but never touches.
- For Linear Inequalities, a dotted-line is used to mean "noninclusive."

Interval, a.k.a **Region** – A set of values in a domain or range, often defined by an inequality statement. The "region" often refers to a place (region) on a number-line or graph coinciding with a set or inequality statement. But (suppose a graph is given first), the graphical region may be represented as a set (domain or range), or an inequality statement. Intervals/Regions are often broken-up parts of a "bigger picture" that may be each individually tested, to see if it/they qualify as being (a true) part of a "bigger picture" inequality. Intervals/Regions can be expressed in Interval and Set-Builder Notation.

INTERVAL & SET-BUILDER NOTATION

Interval notation uses [,], (,), and ∞, and - ∞. Brackets or parentheses can be drawn on
number-line sketches in place of "open or closed-dots."

[- bracket that represents the left or lower limit, meaning "touches," and "inclusive."
(- parenthesis that represents the left or lower limit, meaning "noninclusive" and
"approaches, and gets infinitely close, but never touches."

] – bracket that represents the right or upper limit, meaning "touches," and "inclusive."
) – parenthesis that represents the right or upper limit, meaning "noninclusive" and
"approaches, and gets infinitely close, but never touches."

Note: Whenever a limit is ∞ or negative infinity, a parenthesis is always used. This is to suggest that you never reach infinity, and therefore it is always noninclusive.

Set-builder notation looks like {x| defined region or set of solutions that belong}
In Words, you would read it as: "The set of all elements of x, such that… (then you would read or say the solutions, or inequality statements, or the individual elements defined)."
The "|" is spoken as "such that." Since these sets are usually specifically in regards to values of x, you might also read such a set as:
"The set of all values of x such that… (and read the rest as it is written)."

This notation only uses inequality symbols to indicate one side of a limit, and leaves the "infinity" or "negative infinity" side to be assumed. Both sides (the minimum and maximum) of a set or graph might be shown in consolidated form.

AND vs. OR

The words "and" and "or" are a part of our everyday language, and they are also often used in math. When used in math, it is important that we use them in their correct context, with attention to the detail in their definitions. They may be used slightly differently in the mathematical sense than in the way they are used in casual conversation. "And" and "or" are used in logic, sets, inequalities, and graphing, most notably when graphing sets on number-lines, or vice-versa.

AND
The symbol for "and" is "∩" which stands for "Intersection." It might also be spoken as "intersects with." Think about what "intersection" means… it means "where they cross-over" or "overlap." This is exactly what "and" means.

"And" means "*Only* the values (points) from each set (equation/inequality/graph) which *overlap*. In other words, they *must overlap*.

If you are given two inequalities with "and," and instructed to graph the solution, you should:
1. Graph each inequality separately, on number-lines;
2. Then look to see where they overlap;
3. Then draw a new number-line clearly indicating only the region where the overlap occurs.

See "Inclusive vs. Noninclusive" for the proper graphing symbols.

This is in sharp contrast with the meaning of "or."

OR
The symbol for "or" is "U" which (conveniently) stands for "union" or, as spoken "in union with." The words "or" and "in union with" can be used synonymously.

We often take "or" to mean "*only* one or the other," but in math, it should be thought of as "and/or" or "either/or." More specifically, in regards to *graphing points of a set on a number-line*, "or" should be thought of as the x-values from *both* or *all* appropriate sets. The points or values in the sets *are not required to overlap*, however they may overlap as well. Associate "or" with "all values, whether they overlap or not." So…

If you are given two inequalities with "or," and instructed to graph the solution, you should

1. Graph each inequality (you could do it on one number-line right away)

... and you're basically done; you should sketch a graph with one line encompassing all possible points.

-   Remember, the solution is "all possible points;" they don't have to overlap (although they can)
-   Be sure to use the proper Inclusive/Noninclusive symbols

"And" and "Or" play important roles in: Constraints.

SOLVING (SIMPLE) INEQUALITIES

Inequalities are not "equations" because the quantity on the left side does not equal the quantity on the right (although, the "or-equal-to" inequalities *may*). These are best categorized as math sentences or "statements." It is proper to refer to one as "the inequality," rather than "the inequality equation," (which is a contradiction), as many people mistakenly do.

There are Simple and Compound Inequalities. These are all in reference to a graph.
- A *simple* inequality is *one* inequality (statement), which can be graphed.
- A *compound* inequality is made of *multiple* simple inequalities reflected on the same graph.
    - Two inequalities can be presented in consolidated form.
    - The solution to multiple inequalities can be given in consolidated form.
- Whether an inequality is "simple" or "compound" has nothing to do with the degree (highest power) of the variables involved.

"Solving" an inequality means "simplifying" the inequality statement. The simplified form reflects the region on a graph corresponding to the simplified inequality statement.

For first-degree statements (just x to the unwritten power of 1), your "solution" is one statement in the form of an inequality statement. Often a graph is expected to be sketched to supplement the statement. There are two types of first-degree (linear) inequalities:
- Those with one variable (usually x or y).
    - As "equations," you know these as horizontal or vertical lines.
    - Like the "equation" form, they will have a slope of zero, or no slope.
- Those with two variables (x and y)
    - which, as equations, you know as *linear equations*[*].

First-degree inequalities of *one variable* may be graphed on a number-line, but can be graphed on a 2-Dimensional graph as well, as they need to be when part of a System of Linear Inequalities.

Linear Inequalities of two variables are sketched using a solid or dotted-line and a shaded region, and can only be sketched on a 2-Dimenisonal graph because they have a slope.

For inequalities of degrees of 2 and higher, the "solutions" are still inequality statements, however, there are more steps involved. This is further explained in Solving A Quadratic Inequality.

Solving inequalities are completed in very much the same way as normal equations. The goal is to isolate the variable of interest (usually on the left) with regards to the other term(s) on the other (right) side of the symbol, using mathematical operations and manipulations. There is only one major difference, which is:

When you multiply or divide both sides by a negative number, you must change the direction of the inequality symbol. Note: You only change the direction of the *symbol*. Leave the numbers and variables in place.

There are two ways that multiple inequalities may be given:
- Separate: Each with one inequality symbol and terms on the left and the right, or
- Consolidated: With two symbols, and terms in three places: (the left, the middle – between the two inequality symbols, and the right).

The type with a "left, middle, and right" are a consolidated inequality, which either came from two separate inequality statements (consolidated into one statement, possibly to indicate they are on the same graph), or can be broken up into two separate statements.

Example:
Separate or broken up form: $x > -7$, $x > 3$

Consolidated form: $-7 < x > 3$

You will often be asked to graph your result(s) on a number-line. When doing so, be sure to use the proper inclusive or noninclusive symbols.

You will apply these instructions later, for Solving Compound Inequalities.

SOLVING ABSOLUTE VALUE EQUATIONS

The solutions to absolute value equations are the x-values, or x-intercepts, just as they are for solving non-absolute value equations. The following is a **procedure**:

0. Simplify inside the absolute value group, if necessary.
1. Isolate the whole absolute value group on one side (the left) of the equation.
   - You will then make two equations: a "same" (positive) version and an "opposite" (negative) version...
2. Remove the bars "| |" and set up:
   - The **same** version equation by keeping the expression from inside the absolute value group and the expression on the other side as they originally were, and
   - The **opposite** version equation by keeping the expression from inside the absolute value group the same (on the left), but *changing all signs* in the expression on the other (right) side.
3. Solve each of the two equations separately using algebra techniques, and you will get two solutions (numbers).
   - You may report your answers together, with a comma between them, or circle them separately.
4. You may be told to graph the equation (or inequality) using your solutions (x-intercepts). You can graph the equation like you graph any (non-absolute value) equation[*].
   - Start by making a table of points. Unlike making a table of points for a non-absolute value linear equation, you should make a table similarly to graphing a parabola, because it has a vertex (or vertices).
   - Absolute value graphs (of first degree/linear equations) will make a V-shape or upside-down-V-shape, with one vertex.
     - o The x-values chosen (to find corresponding y-values) should be a few units lesser and greater than the vertex in order to show both sides of the "V" graph with respect to its vertical axis of symmetry.
   - Absolute value graphs of second degree equations may either take the shape of a parabola, or a curved W, or a curved M, depending on if and where the parabola touches/crosses the x-axis.

Look at the following **example** in which the original absolute value equation is:

$$|x + 2| - 6 = 5$$

0. The inside contents of the absolute value group are already simplified.
1. Isolate the absolute value group by adding 6 to both sides:
    $$|x + 2| - 6 + \mathbf{6} = 5 + \mathbf{6}$$

    which becomes:
    $$|x + 2| = 11$$

2. Remove the absolute value bars and set up the *same* and *opposite* versions.

Same Version: $x + 2 = 11$

Opposite Version: $x + 2 = -11$

Notice the "x + 2" stays the same in both. After subtracting 2 from both sides in both equations, the solutions are:

$x = 9$ and $x = -13$, respectively.

SOLVING ABSOLUTE VALUE EQUATIONS w/ TWO ABSOLUTE VALUE GROUPS

Procedure:
0. Simplify inside each absolute value group first, if necessary.
1. Use the proper algebraic techniques to position each absolute value group on opposite sides of the equal sign (before removing their bars).
    *   You will proceed to make two equations: a *same* version, and an *opposite* version. Upon doing so, apply…

2a. "Keep, Keep" for the *same* version, meaning:
Keep the contents (and signs) of the left group the same, after removing the "bars."
Keep all the plus/minus signs from the right group the same, after removing the bars.

2b. "Keep, Change" for the *opposite* version, meaning:
Keep the contents (and signs) of the left group the same, after removing the bars.
Change ALL the plus/minus signs in the right group, to their opposites, after removing the bars.

Note: You need to keep the contents of the absolute value groups on *one side* of the equal sign *the same* in both of the two new equations. It doesn't really matter which of the two sides you keep the same, however for consistency purposes, I recommend keeping the left the same.

3. Simplify the equations from Step 2 (a & b) to get the solutions for x.

Example:

Suppose you are given:

$$|2(x + 3)| = |x + 6|$$

0. Simplify inside the absolute value groups, if possible. Here, the left group can be simplified by distributing the 2 through the parentheses, becoming

$$|2x + 6| = |x + 6|$$

1. The absolute value groups are already on opposite sides of the equation.
This will now be split into two equations...

2a. The Same Version: apply "Keep, Keep":
    $$2x + 6 = x + 6$$

2b. The Opposite Version: apply "Keep, Change":
    $$2x + 6 = -x - 6$$

3. After simplifying, the solutions from the same and opposite versions are:
    $$x = 0 \text{ and } x = -4, \text{ respectively.}$$

SOLVING & GRAPHING A COMPOUND INEQUALITY

A *compound inequality* is a few (usually two) simple inequalities on the same (line) graph.

Procedure:
1. Simplify both inequalities, individually, (so x is isolated on the left in each).

2. Do the simple inequalities have "And" or "Or"? Follow their respective instructions.
   - If no "And" or "Or" are given, follow the Assumptions for "And" & "Or."

3. Make a final, consolidated graph (number-line), according to "and" & "or," and using the proper inclusive vs. noninclusive symbols.

4. Write a consolidated math statement describing the consolidated graph (this may be optional).
You might also give your answer in Interval or Set Builder Notation.

SOLVING & GRAPHING ABSOLUTE VALUE INEQUALITIES

Procedure:
0. Simplify the Absolute Value Inequality: Simplify inside the absolute value group, then isolate the group, if necessary.

1. Go right to setting up the two inequalities (a *same* and *opposite*). (Since you will be setting up two inequalities, you will now technically have a compound inequality).

In short, you might choose to remember:
1a. **"Keep, Keep, Keep"**
1b. **"Keep, Change, Change"**

1a. For Inequality 1 (the Same Version), use **Keep, Keep, Keep**, meaning:
Keep (rewrite) the contents (and signs) of the left group the same, after removing the "bars."
Keep (rewrite) the inequality symbol the same.
Keep (rewrite) the signs from the right group the same.

1b. For Inequality 2 (the Opposite Version), use **Keep, Change, Change**, meaning:
Keep (rewrite) the contents (and signs) of the left group the same, after removing the bars.
Change the inequality symbol to the opposite-direction symbol.
Change ALL the plus/minus signs, in the right group, to their opposites.

2. Now that you have both inequalities set up, finish by following the Procedure for Solving and Graphing a Compound Inequality, and the Assumptions for "And" & "Or", which explain what the graphs will look like.

Example:

Suppose you are given:

$|-2x - 10| > 8$

0. The original inequality statement is simplified: The absolute value group is isolated and the contents of the absolute value group are already simplified.

    This will now be split into two inequalities…

1a. The Same Version: Apply *Keep, Keep, Keep*
$-2x - 10 > 8$

1b. The Opposite Version: Apply *Keep, Change* (to <), *Change* (to negative 8)
$-2x - 10 < -8$

1a (continued). In the *Same Version*, add 10 to both sides to isolate the "-2x"
$-2x - 10 + 10 > 8 + 10$

$-2x > 18$

Divide both sides by -2. Remember that when dividing by a negative in an inequality, you must reverse the direction of the inequality symbol, here, from "greater-than" to "less than."
$x < -9$

1b (continued). In the *Opposite Version*, add 10 to both sides to isolate the "-2x"
$-2x - 10 + 10 < -8 + 10$

$-2x < 2$

Divide both sides by -2. Again, we are dividing an inequality by a negative, so reverse the direction of the inequality symbol, here, from "less-than" to "greater-than."

$x > -1$

Because the original inequality was "greater-than," we assume to give the "or" version answer:

$x < -9$ or $x > -1$

In consolidated form: $-9 > x > -1$
In *set-builder notation:* $\{x \mid -9 > x > -1\}$
In *interval notation:* $(-\infty, -9) \cup (-1, \infty)$

ASSUMPTIONS for "AND" & "OR"

For Simple Inequalities as part of Complex Inequalities where "and" or "or" are not defined:
For < and ≤, assume to answer for "and."
For > and ≥, assume to answer for "or."

Some books can be ambiguous about how they want you to treat (solutions to) compound inequalities when "And," "Or," or their symbols, are not written. They may want you to make the following assumptions:

- If the number-lines *have a region of overlap*, answer with one number-line showing only the overlap, and assume it to be "And (∩)." This is often the case when a region *between* the points of the lines of the two inequalities overlaps.
  - If the "Or" version is answered for a compound inequality of this sort, the answer would be a number-line with a line extending over the whole thing, from negative infinity to positive infinity (meaning: All real numbers), because it would include all points that do and do not overlap.
- If the number-lines *do not* have a region of overlap, answer with one number-line showing both lines, and assume it to be "Or (∪)." This is often the case when the two lines from the inequalities extend in opposite directions of each other, leaving an empty region between them.
  - If the "And" version is answered for a compound inequality of this sort, the answer would be an empty number line (empty set, a.k.a. null-set), because there would be no overlap.

When in doubt, you can at least show both individual graphs, and you can ask your instructor for further clarification.

Assumptions for Quadratic Inequalities:

Positive Parabolas:
For $<$ and $\leq$, assume to answer for "and."
For $>$ and $\geq$, assume to answer for "or."

Negative Parabolas:
For $>$ and $\geq$, assume to answer for "and."
For $<$ and $\leq$, assume to answer for "or."

UNION vs. INTERSECTION

Compare the "solutions" in the following two statements:
$x > 3 \cup x \geq 7$ vs. $x > 3 \cap x \geq 7$

This example is here to highlight and compare examples of the union vs. intersection of two simple inequalities which both "point" in the same direction.

The solution for $\cup$ is: $x > 3$ because, since *union* means all possible points, whether they overlap or not, the simple inequality overlaps the entire domain of $x \geq 7$ and also includes the points between (greater than, not including) 3 and 7. In interval notation, the solution is $(3, \infty)$.

The solution for $\cap$ is: $x \geq 7$ because the portion between (greater than, not including) 3 and 7 does not intersect (overlap), however from (including) 7 up through infinity does overlap. In interval notation, the solution is $[7, \infty)$.

FUNCTIONS

A **function** is an equation that:
- directly represents a graph in which
- no two points have the same x-value,
- but can have the same y-values.

From a graphical standpoint, a function does not cross through any x-value more than once.
- This can be tested with the **vertical line test**.

A function will be written as: $f(x) =$

It is spoken as "f of x" meaning the function is in regards to x-values, and x-values are the *independent variables*. A function can be in regards to other variables, too, as you might explore in some word problems, or calculus. "f(x)" can be thought of as being in place of "y" because values of y are dependent on values of x, which is why "y" is generally considered the universal *dependent variable*.

Some examples of *functions*, by definition, are:
- Linear equations (except for vertical lines)
- Horizontal lines
- (Vertical) Parabolas (Quadratics)
- Cubics
- Exponentials, Logs & Natural Logs
- a half-sideways parabola $\left(y = \sqrt{x}\right)$

Functions, by definition, *can have* more than one y-value associated with a single x-value. A parabola fits this description – every x-value (except for the vertex) has two possible y-values associated with it. Think of the x-intercepts (if the parabola crosses the x-axis)... the y-value for the x-intercepts are both "0".

Some examples of graphs that are *not* functions are:
- Sideways (horizontal) parabolas ($y^2 = x$)
- Circles
- Ellipses
- Vertical lines
- An upright S or Z shaped curve

A function may pass the vertical-line test… but may or may not fail the horizontal line test…

A **One-to-One Function** is a function in which no points share the same y-value… each y-value occurs only once. Graphically, this can be tested with the **horizontal line test**. If a horizontal line will intersect the line of a graph more than once at any place in the graph, the graph *fails* the horizontal line test, and thus *is not* a one-to-one function.

The following common graphs *are* one-to-one functions:
- Diagonal lines (linear equations)
- Vertical lines
- Exponential graphs (including base "e")
- Graphs of logs & natural logs
- The (parent) cubic function: $f(x) = x^3$

The following common graphs *are not* one-to-one functions:
- Horizontal lines $(y = \#)$
- Absolute value graphs (like V or W shape)
- Parabolas (quadratic equations)
- Certain cubic equations
- Ellipses, including Circles (which aren't functions anyway)

Why do you need to know if a function is one-to-one? Because *only one-to-one functions can have an inverse functions*, and logarithms are all based on being inverse functions of exponential functions.

An **Inverse Function** is a function in which the x-values and y-values are switched for each point. Only one-to-one functions can have inverse functions. Graphically, the inverse function is a reflection of the original function at the (diagonal, often dotted) line
$y = x$.

TREND OF A GRAPH

There is an easy way to tell the trend of any graph. Start by asking yourself:

What happens vertically to the graph *as the graph moves to the right* (as x approaches positive infinity)?

If the answer is: It goes up, then the graph is considered: *Increasing* (a positive trend).

If the answer is: It goes down, then the graph is considered: *Decreasing* (a negative trend).

Although this may seem obvious, this is to help you properly define a graph as increasing or decreasing.

There are clues in equations which will tell the trend of a graph.
For linear equations, in slope-intercept form[*], the sign of the slope, m, tells you the trend.
- If it is positive, there is a positive trend, and it goes up and to the right.
- If it is negative, there is a negative trend, and it goes down and to the right.

For quadratic equations, the sign of "a", the leading coefficient, tells the trend of the graph.
- If it is positive, there is a positive trend (opens upwards like a U-shape).
- If it is negative, there is a negative trend (opens downwards like an upside-down U).

GRAPHING LINEAR INEQUALITIES (in two variables)

The graphical form of a Linear *Equation* is a line on a graph. The graphical form of a Linear *Inequality* is a half-shaded graph, in which a *line* separates the shaded side from the non-shaded side. You test a test-point to determine which side is shaded.

Linear inequalities (in two variables) are very closely related to *linear equations*; the only differences between their graphs are that:
- the graph of the linear inequality is shaded on one side or the other of the line, and
- the line may or may not be dotted to show inclusiveness/noninclusiveness.

The following is a **procedure** for graphing linear inequalities in two variables:

0. Simplify the inequality: Solve for (isolate) y. If it is a function, the f(x) will be isolated already.
   - You will need this inequality a few steps later (Steps 4 - 6)

1. Rewrite the simplified inequality as an equation, by replacing the inequality symbol with an = sign.

2. Plot the line. There are two ways to do this:
   - Make a table of 3 points*, then graph; or,
   - The shortcut: Use the y-intercept and slope to make the line.

3. Is the inequality inclusive or noninclusive?
   - If Inclusive, draw a solid line through the points.
   - If Noninclusive, draw a dotted line through the points.

4. Look at the graph carefully and choose a *test-point* which is not *on* the line. If possible, choose (0, 0), because it will help you simplify the quickest. If the line goes through (0, 0), chose another point near the origin that the line doesn't go through, like (0, 1), (1, 0), or (1, 1).

5. *Test* the test-point in the simplified *inequality* you have *from the beginning*, by substituting the x-value in for x and the y-value in for y [or in place of f(x)].

6. Simplify, and test the truth of the statement.
   - If the statement reads to be *true*, shade the side of the graph that test point is in.
   - If the statement turns reads to be *false*, shade the *opposite side* of the line where the point was tested.

Example:

Suppose you are given y – 2x > 1

0. Simplify. Isolate the y by subtracting 2x from both sides.

   y > 2x + 1

1. Rewrite the inequality as an equation, replacing the ">" with "=".

   y = 2x + 1 (This is now in slope-intercept form).

2. Take the steps to graph it. Use the y-intercept, giving you one point: (0, 1). You could use the slope 2 (think of it as 2 over 1) to find another point by starting at the y-intercept and counting up 2 units, then over to the right 1 unit. Or, find two more points using substitution to make a table.

   If x = 1,

   y = 2(1) + 1, so y = 3 → (1, 3), and

   if x = 2,

   y = 2(2) + 1, so y = 5 → (2, 5).

x	y
0	1
1	3
2	5

3. Graph these three points. Sketch a *dotted line* since ">" is noninclusive.

4. Is the point (0, 0) on the sketched line? No, so use (0, 0) as the test-point.

5. Substitute 0 in for y and x in the original inequality (you could substitute it into the original, or the simplified inequality; the outcome will be the same).

   $0 > 2(0) + 1$ which simplifies to

   $0 > 1$, which is a false statement, because zero is not greater than one.

6. Since it is false, shade the side of the graph *opposite* of the side that has (0, 0).

SYSTEMS OF LINEAR INEQUALITIES

A *system* of linear inequalities is nothing more than multiple, overlapping, individual linear inequalities (the linear inequality statements may have one or two variables). That being said, there is little to the procedure for *solving* a system of linear equations.

The solution is: the region where the shaded-regions of all the linear inequalities in the system overlap, *if* they overlap. (The solution *is not a single point*, as is the case for Solving Systems of Linear Equations*).

If they *do* overlap, be sure to highlight that region.
If they do *not* overlap, there is *no solution*, and you should state that, but you should still show the graph with each properly shaded region.

In short, you are graphing each inequality (line, shaded to one side), on the same graph.

The **procedure** is really just a matter of following the Procedure for Graphing Linear Inequalities (in two variables) for each inequality, on the same graph, and identifying the overlapping shaded region.

Note: You might have two *or more* linear inequalities, like five inequalities. No matter how many inequalities are given in the system, just look for the shaded-region where they ALL overlap.

Keep in mind that the inequalities in a system could include *inequalities of one variable*, which would either be:
- a horizontal line shaded above or below it, or
- a vertical line shaded to the left or right of it.

This could include the x-axis and/or y-axis, shaded to one side, as part of the system of inequalities. This might be confusing at first because inequalities of one variable are usually shown on a number-line (a one-dimensional graph), but when they are part of a system, they are shown on a two-dimensional graph.

A LINE PERPENDICULAR TO A LINE AT A GIVEN POINT

If you are given a linear equation in standard form and told to write an equation in standard form that is perpendicular to that line and passes through a given point, use the following **procedure**:

1. Convert the given equation in standard form into slope-intercept form.
2. Extract the slope (m) from slope-intercept the equation you just converted.
3. Do the proper slope conversion for a perpendicular by making it the opposite reciprocal.
4. Using the given point and the slope you just found, substitute these into the point-slope formula.
5. Rearrange this equation into standard form.

Note: If you are originally given only 2 points instead of an equation (and a point), you can use those two given points to get the answer. The difference is: Instead of doing Steps 1 & 2, listed above, you will need to find the slope of the two given points using the slope equation[*]. Then proceed with Step 3, and use either one of the points originally given, for Step 4.

DIVIDING POLYNOMIALS USING LONG DIVISION

The following is a step-by-step procedure for dividing polynomials using long division.

0. Setup:
0a. Identify the divisor and dividend and put them in their proper places[*].
0b. Be sure all terms are in descending order.
0c. Fill in all missing powers of x with a placeholder in both the divisor and dividend, if needed. A placeholder is x written to the missing power, with a coefficient of 0. The reason for the placeholder is so that like-terms can be properly added or subtracted, as in Step 4.

Procedure:
1. Ask yourself: How many times does the leading (left-most) term of the divisor go into the leading (left-most) term of the dividend? Write the answer above the leading term of the dividend.

2. Take the number you just wrote above and multiply it (first) by the leading term of the divisor, then continue multiplying it by all other terms of the divisor, in successive order. Write your answer(s) below their like-terms of the dividend. (The first product you write should be the same number as directly above it, which, in the first run-through, will be the leading term of the dividend).

3. Of the terms you just wrote from multiplying, clearly change all the signs to their opposites.

4. Perform the arithmetic (whether it's addition or subtraction); the left terms should always cancel to 0 (if they don't, you made a mistake).

5. Bring down the next term (including the sign) from the dividend so it is next to the difference from Step 4.

6. Repeat the procedure (Steps 1 through 5) from above for each term of the dividend, first, by asking how many times the leading term of the divisor goes into the left-most term you found after bringing down a term.

Last Step: After repeating the steps until there are no more terms from the dividend to bring down, there are two ways the problem will conclude. If the last terms cancel to "0", the problem is complete. If the last step yields a non-zero number, this is the remainder, but you must assign the remainder properly. In your quotient section, add the fraction of your remaining term over the divisor. This fraction is technically the remainder.

SPECIAL CASE: THE SUM OR DIFFERENCE OF TWO CUBES

The sum or difference of two cubes is considered a "special case" polynomial. Before looking at the procedure, it's important to acknowledge that this is distinctly different than "the sum of two squares" and "the difference of two squares," mainly because the sum of two squares is prime, but the sum of two cubes can be factored. The sum or difference of two cubes are factored according to the following formula:

Sum: $\qquad a^3 + b^3 = (a + b)(a^2 - ab + b^2)$

Difference: $a^3 - b^3 = (a - b)(a^2 + ab + b^2)$

In short, whether the original binomial is the sum or difference of two cubes, they are always factored into two polynomial factors: a binomial times a trinomial.

In Word form: Original binomial = (binomial factor)(trinomial factor)

In the formulas shown above, the "a" and "b" represent whole terms, which may be a combination of coefficients and variables. When these are in terms of x and y, with no coefficients, it will look closer to:

Sum: $\qquad x^3 + y^3 = (x + y)(x^2 - xy + y^2)$

Difference: $x^3 - y^3 = (x + y)(x^2 + xy + y^2)$

When coefficients are involved, you might think of it as:

Sum: $\qquad \#^3x^3 + \#^3y^3 = (\#x + \#y)(\#^2x^2 - \#xy + \#^2y^2)$

Difference: $\#^3x^3 - \#^3y^3 = (\#x + \#y)(\#^2x^2 + \#xy + \#^2y^2)$

Procedure for factoring the Sum or Difference of Two Cubes

The **procedure** makes reference to the original binomial as:

$$\#^3x^3 \qquad\qquad \pm \qquad\qquad \#^3y^3$$

The first cube, the sign between, and the last cube

Regardless of whether you have the sum or difference of two cubes to start, the procedure for factoring them is the same.

0. Set your original binomial equal to an empty binomial and trinomial, then fill them in as you go; leave yourself enough space.

$$\#^3x^3 \pm \#^3y^3 = (\quad\ )(\qquad\qquad)$$

Fill in the binomial factor first. Then, make the trinomial factor based off of the binomial you just made.

1. The binomial factor will be made of the cube roots of the first and last terms of the original expression.
   - The sign of the original expression is the same as the sign in the binomial factor.

2. The terms in the trinomial factor can be made in reference to the binomial factor you made in Step 1:
   2a. The first term is the square of the first term of the binomial factor.
   2b. The middle term is the product of the first and last term of the binomial factor... and it always carries the opposite sign of the sign between the two cubes of the original expression (and the binomial factor).
   2c. The last term in the trinomial always carries a positive sign.
   2d. The last term of the trinomial is the square of the last term of the binomial factor.

3. Can it be factored further?
   - If the leading term of the trinomial factor is only to the power of two, then this trinomial is prime and can't be factored any further. However, if the power of the variable of the leading term is higher than two (it can only ever be a multiple of two), it must be re-evaluated to see if it can be further factored. An example of when it can be further factored is shown in the next section.

- Likewise, the binomial factor should be analyzed as well, to check if it can be further factored. The difference of two squares should be factored. The sum of two squares is prime. If you still have the sum or difference of two cubes, which you may if the powers of the original sum or difference of two cubes were multiples of 3, repeat the process for factoring.

The other special cases:
- The Difference of Two Squares,
- The Sum of Two Squares, and
- Perfect Square Trinomials

are discussed in more detail in the first *ALGEBRA IN WORDS: A Guide of Hints, Strategies and Simple Explanations*.

Also: "A Binomial Cubed" is sometimes considered a special case. Binomials raised to any power can be expanded using a formula for "Binomial Expansion," not covered in this book.

Can a Perfect Square also be a Perfect Cube?

Yes, certain higher powers can be both perfect squares and perfect cubes, if they are multiples of both 2 and 3. The best example is a power of 6, and all powers which are multiples of 6. If you encounter the difference of two powers of 6...

Factor them as the *difference of two squares* first, to get conjugate-pair binomials.
The terms in those conjugate binomials will be in powers of 3 (or multiples of 3). In other words, the conjugate binomials will include one factor as *the difference of two cubes*, and the other factor as *the sum of two cubes*. Proceed to factor them according to the special case for the sum or difference of two cubes.

Take the **example**: Completely factor:
$64x^6 - 729y^{12}$

Notice that:
- the coefficient $64 = 2^6$ and is a perfect square (8^2) and cube (4^3);
- the coefficient $729 = 3^6$ and is a perfect square (27^2) and cube (9^2);
- the exponent 6 is a multiple of 2 and 3, either as x^{2^3} or x^{3^2}; and
- the exponent 12 is a multiple of 6, and therefore also of 2 and 3.

First, factor this *as the difference of two squares* into conjugate-pair binomials:
$(8x^3 - 27y^6)(8x^3 + 27y^6)$

which are now the difference of two cubes and the sum of two cubes, respectively. Factoring them as such, they factor into:
$(2x - 3y^2)(4x^2 + 6xy^2 + 9y^4)(2x + 3y^2)(4x^2 - 6xy^2 + 9y^4)$

which cannot be factored any further.

But now, let's see how this would go if you factored $64x^6 - 729y^{12}$ in reverse order (first as the difference of two cubes). You would get:

$(4x^2 - 9y^4)(16x^4 + 36x^2y^4 + 81y^8)(4x^2 + 9y^4)(16x^4 - 36x^2y^4 + 81y^8)$

It is obvious that the first (left) group is the difference of two squares, which could be factored into:

$(2x - 3y^2)(2x + 3y^2)$... but where do you go from there, with regards to factoring the other factors? Can it be done? Yes, but the truth is... it just becomes a long, tedious, and unnecessarily complicated process. It is also not easy to look at the remaining factors to know they can be reorganized and factored into simpler factors. If you wrote:

$$(2x - 3y^2)(2x + 3y^2)(16x^4 + 36x^2y^4 + 81y^8)(4x^2 + 9y^4)(16x^4 - 36x^2y^4 + 81y^8)$$

as your answer on a test, you would get it wrong.

This section is just to prove two points:
1. Perfect squares can also be perfect cubes, and
2. When you are factoring a binomial which is the difference of both two squares and two cubes, treat it as the *difference of two squares*, and start by factoring them into *conjugate binomials*. Then factor those factors as the *sum and difference of two cubes*.

Also, it is important to point out that this only applies to the *difference* of two terms, not the sum, because the *sum of two squares* cannot be factored into real conjugate pair factors. Therefore, the sum of terms which are both squares and cubes still could (and should) be treated and factored, first, as cubes.

DIFFERENT WAYS TO SIMPLIFY A RATIONAL EXPRESSION

$$\left(\frac{ab^{-3}cd^5f^{-4}}{8a^2b^{-2}cd^3}\right)^{-3}$$

At some point, you will be required to simplify a rational expression of this type, involving coefficients and variables raised to positive and negative powers in both the numerator and denominator, all taken to a negative exponent. In a problem like this, you must recall the principles of:

- Division of common base variables
- Negative exponents
- Distributing an exponent, and
- Powers of powers

There are three acceptable ways to begin simplifying a rational expression of this type, while still following the order of operations. They are:
1. Simplifying inside the parentheses
2. Distributing the outer exponent
3. Changing the outer exponent from negative to positive

We will examine each way, in detail, starting on the next page.

Starting Approach #1: Simplifying inside the parentheses.
Start with:

$$\left(\frac{ab^{-3}cd^5f^{-4}}{8a^2b^{-2}cd^3}\right)^{-3}$$

Intermediate Step showing the subtraction of exponents from the numerator and denominator:

$$\left(\frac{a^{1-2=-1}b^{-3-(-2)=-1}c^{1-1=0}d^{5-3=2}f^{-4}}{8}\right)^{-3}$$

which becomes:

$$\left(\frac{a^{-1}b^{-1}d^2}{8f^4}\right)^{-3}$$

Notice the c variable cancelled out to 1 because it was simplified to the power of zero.
Adjusting for the negative exponents of a and b, they are moved to the denominator:

$$\left(\frac{d^2}{8abf^4}\right)^{-3}$$

Intermediate Step, showing the distribution of the outer exponent, multiplied times the inner exponents:

$$\left(\frac{d^{2(-3)}}{8^{-3}a^{-3}b^{-3}f^{4(-3)}}\right)$$

which becomes:

$$\left(\frac{d^{-6}}{8^{-3}a^{-3}b^{-3}f^{-12}}\right)$$

All bases are moved to the opposite part of the fraction due to the negative sign in their exponents; then 8 is cubed, to become the final simplified form:

$$\left(\frac{512a^3b^3f^{12}}{d^6}\right)$$

This approach took about 4 steps.

Starting Approach #2: Distributing the outer exponent.
Start with:

$$\left(\frac{ab^{-3}cd^5f^{-4}}{8a^2b^{-2}cd^3}\right)^{-3}$$

Intermediate Step, showing the distribution of the outer exponent, multiplied times the inner exponents:

$$\left(\frac{a^{-3}b^{-3(-3)}c^{-3}d^{5(-3)}f^{-4(-3)}}{8^{-3}a^{2(-3)}b^{-2(-3)}c^{-3}d^{3(-3)}}\right)$$

It becomes:

$$\left(\frac{a^{-3}b^9c^{-3}d^{-15}f^{12}}{8^{-3}a^{-6}b^6c^{-3}d^{-9}}\right)$$

At this point, there are two directions you can go. You can either adjust for the negative exponents, or you can divide (and consolidate) variables of a common base using subtraction. I'm going to divide and consolidate variables in the intermediate step:

$$\left(\frac{a^{-3-(-6)}b^{9-6}c^{-3-(-3)}d^{-15-(-9)}f^{12}}{8^{-3}}\right)$$

which becomes:

$$\left(\frac{a^3b^3d^{-6}f^{12}}{8^{-3}}\right)$$

Then adjust for negative exponents:

$$\left(\frac{8^3a^3b^3f^{12}}{d^6}\right)$$

Cube 8, and the final simplified form is

$$\left(\frac{512a^3b^3f^{12}}{d^6}\right)$$

This approach took about 4 steps

Starting Approach #3: Changing the outer exponent from negative to positive. Start with:

$$\left(\frac{ab^{-3}cd^5f^{-4}}{8a^2b^{-2}cd^3}\right)^{-3}$$

To change the outer exponent "-3" positive, flip the fraction in the parentheses to the reciprocal:

$$\left(\frac{8a^2b^{-2}cd^3}{ab^{-3}cd^5f^{-4}}\right)^3$$

Keep in mind, this only allowed the outer exponent -3 to change to positive; do not change the signs of the exponents of the factors inside the parentheses (yet), even though they changed position within the fraction. If you change them (again, right now), it undoes the change from the exponent "-3" to "3".

From here, you can proceed in two ways:
- Simplify inside the parentheses, or
- Distribute the outer exponent "3".

I'm going to distribute the outer exponent 3, shown in the next intermediate step:

$$\left(\frac{8^3a^{2(3)}b^{-2(3)}c^3d^{3(3)}}{a^3b^{-3(3)}c^3d^{5(3)}f^{-4(3)}}\right)$$

Simplify: Multiply the exponents and cancel out the c^3s:

$$\left(\frac{8^3a^6b^{-6}d^9}{a^3b^{-9}d^{15}f^{-12}}\right)$$

Intermediate Step showing the subtraction of exponents from the numerator and denominator:

$$\left(\frac{8^3a^{6-3}b^{-6-(-9)}d^{9-15}}{f^{-12}}\right)$$

which becomes:

$$\left(\frac{8^3a^3b^3d^{-6}}{f^{-12}}\right)$$

Move the variables with negative exponents to the opposite part of the fraction, making their exponents positive, and expand 8^3, to get:

$$\left(\frac{512a^3b^3f^{12}}{d^6}\right)$$

This approach took about 4 steps as well.

Discussion of the Three Ways

Looking back at the three ways to begin, which method proved best… or fastest? Looking at it objectively, each approach took about four steps, so neither way proved to be fastest. If you trace back, you might notice that each approach actually used the same steps (principles), in various order:

- Distributing the outer exponent of -3 to each factor in the parentheses (by multiplying their exponents)
- Dividing variables of a common base by subtracting the exponent in the denominator from the exponent in the numerator
- Removing negative signs from exponents by moving their base to the opposite part of the fraction, then making the exponent positive
  - In the case of the outer exponent, converting the rational expression in the parentheses to the reciprocal (makes the "-3" become positive "3").
- Converting 8^3 to 512

Simplification problems like this are tedious and require a lot of focus toward many details. The main thing is to realize that there can be multiple, acceptable approaches to the same problem, to attain the same result. I say this because some students see the initial problem and instantly freeze at the fear that they are unsure at what specifically to do at first. That is okay; just start somewhere, and work your way through, step-by-step. Leave yourself plenty of room on your paper, write each step clearly, writing the intermediate steps if you think it will help, apply the fundamental properties one-at-a-time, and pay close attention to the details. As this section was meant to prove that each approach was equal in ease, you are encouraged to begin with the step you feel most comfortable doing.

SOLVING AN EQUATION with MANY FACTORS for a PARTICULAR VARIABLE

A subsection of this sort is often included when you do rational equations. You are given an equation and asked to solve for a particular variable (or sometimes, a group of factors). The equations given are often actual formulas from the fields of geometry, chemistry, physics, trigonometry, economics, and statistics. Exercises in solving multi-variable equations in terms of other variables prepare you for using them in other classes.

Students are sometimes thrown off by these types of problems for some reason, but they are actually very simple, because they draw off of simple algebraic manipulation principles[*]. The extra challenges might lie in the facts that:

- They may contain more factors and variables than you are used to seeing in more traditional polynomials;
- They may involve a variety of different variables, other than x & y, such as different letters, symbols, possibly Greek letters, capital & lowercase, etc., that may contain numeric or alphabetic subscripts, like m_2 or T_c (which stands for temperature in Celsius);
- They may require taking a root of both sides;
- *Your answer will be in terms of other factors* which may contain variables and number-factors; your answer *will not be a number*, as you may be used to getting in other traditional equations.

You will be told which variable to solve for.
Look at the given problem and locate the variable to solve for.
Now locate the variable in a more general sense… which side is the variable that you are solving-for on? It will likely be clustered among many other factors and variables, and it will likely appear in one (or a combination) of the following common forms:

1. Inside a group of parentheses
2. Wedged amongst a many factors (could be within a fraction)
3. It may appear multiple places (in more than one term)
4. It might appear in the denominator
5. It may appear squared or raised to an exponent

I will walk you through the step(s) to take when the variable appears in one of the described positions, then give a coinciding example.

1. If it is inside a group of parentheses, you may need to distribute the factor in front of the parentheses in order to break the term with the factor you are solving-for out of the parentheses.

2. If it is wedged amongst many factors, follow this instruction carefully: Locate the side of the equation with the variable you are solving-for. Multiply both sides of the equation by the *reciprocal factors* of *all factors except the one you are solving for*. You might solve the problem for the variable in this one step. (However, if the variable you are solving-for is taken to a power, continue with instruction 5.)

3. If the variable you are solving-for appears more than once, possibly in more than one term, you must factor out the variable you are solving-for as a GCF. Move all other terms, if any, to the other side of the equation using addition or subtraction; this will isolate the term consisting of the GCF (you just factored out) and the group of parentheses (the GCF was factored out of). Divide both sides by the group of parentheses. That will likely be the last step needed to solve for your variable of interest.

4. If the variable you are solving-for is in a denominator, you should cross-multiply to move the variable to the numerator. But remember, you can only cross-multiply if there is only one fraction (or term) on each side. If your variable of interest appears in multiple fractions, you might continue by either adding the fractions together (thus consolidating them into one fraction, and allowing you to cross-multiply next), or, you might multiply all terms by the LCD, and continue from there.

5. If the variable you are solving-for is squared or raised to another exponent, you must first isolate the base taken to that exponent, and you might do this using the instructions above. If the variable you are seeking to solve for is squared, solve using the Square Root Property (take the square root of both sides). If the variable is to a higher power, take the root of that power of both sides (put the radical over the term with the power and the entire other side). Note: If you take the root of both sides, and there is a denominator on the other side, you may need to rationalize the denominator to complete the problem.

The following examples are numbered to coincide with the list above.

Example for #1: Solve for b in:

$$A = \frac{h(B + b)}{2}$$

This formula is a form of the equation for the Area of a Trapezoid (from geometry), in which:
- h is height
- b is the length of the top line, and
- B is length of the base

Notice the subtle difference between the capital and lowercase b. Be sure to reflect the proper forms in your work.

Distribute the h through the parenthetical group, becoming:

$$A = \frac{hB + hb}{2}$$

Think of A, on the left, as A-over-1, then cross multiply, to get:
$$2A = hB + hb$$

Move hB to the left by subtracting it from both sides, which becomes:
$$2A - hB = hb$$

Divide both sides by h, which isolates b:
$$\frac{2A - hB}{h} = b$$

The task of solving for b is complete, and this answer is acceptable, however, you could further simplify the left side, by splitting up the fraction,

$$= \frac{2A}{h} - \frac{\cancel{h}B}{\cancel{h}}$$

then cancelling out the h from the right fraction, to get:

$$b = \frac{2A}{h} - B$$

Example for #2: Solve for m_1 in:

$$F = \frac{Gm_1 m_2}{d^2}$$

This is the equation for the Gravitational Force between (the masses of) two (spherical) bodies in the universe (from Newton's law of universal gravitation, in classical mechanics physics), in which:

- F is gravitation force between two bodies
- G is the gravitational constant
- m is the mass of a (spherical) body
- The subscripts 1 and 2, of each m, are there to indicate that they are masses of different entities
- d is the distance between the centers of each spherical body (sometimes r is used in place of d)

Since the variable of interest, m_1, is in the numerator, look at all the other factors on that side. This problem can be solved in one simple step.
Look at the equation in this intermediate step, which just rearranges the order of the factors, to make the variable you are solving for and the factors to move, standout:

$$F = \frac{m_1}{1}\left(\frac{Gm_2}{d^2}\right)$$

Multiply both sides by the reciprocal of all the factors other than m_1, which cancels out all the factors… except m_1, on the right:

$$\frac{Fd^2}{Gm_2} = \frac{Gm_1 \cancel{m_2}}{\cancel{d^2}}\left(\frac{\cancel{d^2}}{\cancel{Gm_2}}\right)$$

And you are done:

$$m_1 = \frac{Fd^2}{Gm_2}$$

Example for #3 & #4: Solve for P in:

$$\frac{A - P}{Pr} = t$$

This formula is a form of the equation for Simple, Accrued Interest (from economics), in which:
- A is total accrued amount (sum of principle plus interest made)
- P is principle (the starting amount being invested)
- r is the interest rate
- t is time period

and the variable you are solving for is both:
- in two places, including
- in the denominator

Think of t as t-over-1, then cross multiply, to move Pr out of the denominator, to get:
$Pr = t(A - P)$

On the right, distribute the t through the parentheses, which will break the P out of the parentheses, and make it available to move, in the next step.
$Pr = tA - tP$

Move tP to the left by adding it to both sides, so both terms containing P are on the same side.

$Pr + \mathbf{tP} = tA - tP + \mathbf{tP}$, so...

$Pr + tP = tA$

Factor P out of both terms on the left:

$P(r + t) = tA$

Divide both sides by the whole parenthetical group $(r + t)$, moving it to the denominator on the right, to get the answer:

$$P = \frac{tA}{(r + t)}$$

Example for #4 & #5

This is, again, using the equation for Gravitational Force between the masses of two spherical bodies in the universe, from the example for #2, but this time, solve for d^2 (first, then for d), which is both:
- in the denominator, and
- a squared term.

$$F = \frac{Gm_1m_2}{d^2}$$

In one step, you can:
- get the d^2 out of the denominator,
- move the F to the (denominator of the) other side, and
- isolate d^2

if you multiply both sides by $\frac{d^2}{F}$

$$\frac{\cancel{F}d^2}{\cancel{F}} = \frac{Gm_1m_2}{\cancel{d^2}}\left(\frac{\cancel{d^2}}{F}\right)$$

which becomes:

$$d^2 = \frac{Gm_1m_2}{F}$$

Now, since d is squared, take the square root of both sides:

$$\sqrt{d^2} = \sqrt{\frac{Gm_1m_2}{F}}$$

and you are done:

$$d = \sqrt{\frac{Gm_1m_2}{F}}$$

CONVERTING RADICALS & RATIONAL EXPONENTS

Radicals can be converted into *rational exponent form*. Rational exponents are a form of exponents which consolidates the *power* and *root* into one fractional exponent… or *ratio*.

Consider the following generic example (one you will not likely see in textbooks),
showing the conversion of radical form to rational exponent form, in which
"p" represents the *power* (either inside or outside a radical), and
"r" represents the *root* of some radical.

$$\sqrt[r]{x^p} = (\sqrt[r]{x})^p = x^{\frac{p}{r}}$$

$$\sqrt[denominator]{x^{numerator}} = (\sqrt[denominator]{x})^{numerator} = x^{\frac{numerator}{denominator}}$$

Notice how the power becomes the numerator, and the root becomes the denominator, of the *rational exponent*. Rational-exponent-form can likewise be converted back to radical form.

This concept gives clarity to three concepts:
- The definition of higher-order roots,
- Simplifying expressions with rational exponents, and
- Evaluating roots with various factors & exponents.

Regarding *the definition of higher-order roots*, consider the case when p = r.

This would cause the numerator and denominator to be the same, and any number over itself equals 1. Thus, any base to the power of 1 equals itself (that base).

Also, any rational exponent can be converted to (or from) decimal form.

Examples:

Convert $\sqrt{x}$ to rational exponent form. This example is chosen because square roots are an unwritten root of "2", and any term is assumed to be raised to an unwritten power of "1". Therefore,

$$\sqrt{x} = x^{\frac{1}{2}}$$

Convert $x^{\frac{-1}{3}}$ to radical form:

$$x^{\frac{-1}{3}} = \frac{1}{\sqrt[3]{x}}$$

In Words: The denominator 3 becomes the cube root, and the negative causes the base x (now, the cube root of x) to be moved to the denominator.

Convert the following to radical form:

$$-\frac{1}{x^{\frac{-3}{4}}}$$

In Words: This is "negative one over x to the negative three-fourths power."

$$-\frac{1}{x^{\frac{-3}{4}}} = -\sqrt[4]{x^3} = -\left(\sqrt[4]{x}\right)^3$$

Comments: As the x-term with the negative rational exponent is originally in the denominator, the negative sign is removed by moving it up to the numerator. Notice that this causes no change to the negative sign (to the left) of the fraction, which remains.

SIMPLIFYING EXPRESSIONS WITH RATIONAL EXPONENTS

Solving (or evaluating, as it may be called) a term or base with a rational exponent may be done in two steps:

1. Apply the root (denominator part of the exponent).
2. Apply the power (numerator part of the exponent).

Notes:
- You don't have to follow that order. You can apply the power first, then the root, however, I believe applying the root first, then the power, is favored because you may encounter more manageable, familiar numbers.
- If the rational exponent is negative, remember to flip the base (to its reciprocal), and remember to bring the (now positive) rational exponent with it (see the example).
- If you are permitted, you might also use your calculator,
  - however, some calculators might only report irrational numbers in decimal form and not radical form.
  - Some calculators have designated buttons for exponents. When inserting a rational exponent, you should put it in parentheses to maintain proper order of operations.

These are shown in the following simple examples...

Example: Evaluate: $\sqrt[3]{8^5}$

Convert to rational exponent form: $8^{\frac{5}{3}}$

The denominator is 3, so take the cube root of 8, which is "2" giving you 2^5. Since the root (denominator) was used, it is removed. The original base "8" is converted to "2" from the previous cube root-step. Now apply the numerator, which is the power of 5:

$2^5 = 32$

Similar Example but with a negative rational exponent:

If you were given $8^{\frac{-5}{3}}$

Get rid of the negative sign in the exponent by moving the entire base with the (now positive rational exponent) to the denominator as:

$$\frac{1}{8^{\frac{5}{3}}}$$

Under 1, take the cube root of 8, giving 2, then raise it to the 5th power, as in the last example, and the answer is:

$$\frac{1}{32}$$

FINDING HIGHER POWERS OF i

By now, you probably know that "i" is a symbol (which stands for "imaginary number") used as a substitute for the square root of -1 (because you can't take the even root of a negative radicand to get a real number).

Sometimes you have to substitute i for the square root of negative one, or sometimes you have to do the reverse. Sometimes, you will encounter higher powers of i.

The following is a list of the four primary, common, lower powers of i:

$i = \sqrt{-1}$

$i^2 = -1$

$i^3 = -1$

$i^4 = +1$

Look again at the list below, showing how the powers of i go in repeating consecutive cycles of four possible simplified answers; a few intermediate steps are shown to give clarity and perspective as to where the solutions come from:

$i^0 = +1$

$i \text{ or } i^1 = \sqrt{-1}$

$i^2 = \left(\sqrt{-1}\right)\left(\sqrt{-1}\right) = \left(\sqrt{-1}\right)^2 = -1$

$i^3 = i^2 i^1 = -1\sqrt{-1} = (-1)(i) = -i \ (\text{or} - \sqrt{-1})$

$i^4 = (i^2)(i^2) = (-1)(-1) = +1$

If you want less to remember, just memorize i and i^2 because the i-cubed and i-to-the-fourth can both be condensed to factors of only i and i^2, respectively, as you can see in the following:

$i^3 = (i)(i^2) = -i$, and

$i^4 = (i^2)(i^2) = +1$

There are a number of ways to "evaluate" (reduce) a higher power of i. Chose the method(s) you feel most comfortable doing. In the following section, I will explain the methods. Although this may seem like a lot to remember at first, it really won't seem that way once you get the hang of it, and the repetitive relationship of the numbers in the exponents. These methods will be demonstrated in the Examples section, where we will examine a bunch of examples.

Methods:
You should always begin by seeing if the exponent is a multiple of 4, because if it is, the answer is "1" and there's no more work to do. If the exponent is not a multiple of 4, use Method 1 if it is even, and Method 2 if it is odd.

Note: If the exponent is even, but not a perfect multiple of 4, the answer can only be "-1". Although you can easily deduce this, you will need to prove it. Method 1 discusses how to prove it, mathematically.

Method 1 is based on the principle of taking a *power to a power* (in which you multiply the exponents) using i^2 as the base. Split up the exponent by dividing it by 2, assigning the 2 to the base i^2 and the quotient as the outer exponent. This is useful because i^2 converts to "-1", and the last step is raising that to a positive or negative power.
$$(i^2)^{\#} = (-1)^{\#}$$

Substitute "-1" in for the base i^2. The remaining factor of the overall exponent will become the outer exponent which should then be applied to the base. Raising the base "-1" to an odd power yields "-1", and raising it to an even power yields positive "1". This method is quick and easy but only works for even exponents. Luckily, any odd number is only one unit away from an even number, which can be easily applied in the next method.

Method 2 is based on the principle of *multiplying factors with a common base* by adding their exponents. If the exponent is odd, break it up into a positive part (using Method 1) times "i" as:
$$(i^2)^{\#}(i) = (-1)^{\#}(i)$$
In short, this will result in an either positive or negative "i". This method is built off of "only needing to memorize i^1 and i^2", because any exponent can be factored in terms of them.

Method 3 is centered around *where the exponent fits in the cycle of 4.* You should always start by seeing if the exponent is a multiple of 4. If it is, the answer is 1, and you're done. You might use this method if the exponent is not a multiple of 4. However, to start this method, you should find the multiple of 4 that is closest to the exponent you were given, as a point of reference, then count a few units up or down until you find the place in the cycle.

Examples:

i^6 6 is not a multiple of 4, and is even. Use Method 1:
Factor 6 into 2 times 3:
$(i^2)^3 = (-1)^3 = -1$

i^7 7 is not a multiple of 4, and is odd. Use Method 2:
$i^{3+4} = (i^3)(i^4) = (-i)(1) = -i$

i^{10} 10 is not a multiple of 4. Use Method 1:
$(i^2)^5 = (-1)^5 = -1$

$i^{12} = 1$
because 12 is a perfect multiple of 4

i^{29} 29 is not a multiple of 4, and is odd. Use Method 2:
$i^{28+1} = i^{28}(i) = (i^2)^{14}(i) = (-1)^{14}(i) = (+1)(i) = i$

i^{38} 38 is not a multiple of 4. Use Method 1:
$(i^2)^{19} = (-1)^{19} = -1$

Using Method 3:
38 is not a perfect multiple of 4.
40 is the next higher multiple of 4
Starting from 40, count backwards 2 units to 38.
Therefore, this can be thought of as
$i^{38} = i^{40-2} = i^2 = -1$

$i^{76} = 1$
because 76 *is* a perfect multiple of 4

i^{97} 97 is not a multiple of 4, and is odd. Use Method 2:
$i^{96+1} = i^{96}(i) = (i^2)^{48}(i) = (-1)^{48}(i) = (+1)(i) = i$

Recap *in Words*:

If the exponent of i is even, it will yield either positive or negative 1.
If the exponent is a perfect multiple of 4, it equals 1. If it is not a perfect multiple of 4, but is even, it equals -1.

Likewise, when you divide the exponent by 2:
- If the quotient is odd, the answer is -1, because negative 1 to any odd power equals negative 1.
- If the original exponent divided by 2 is even, the answer will be positive 1, because "(-1)" raised to any even power equals positive 1.

If the exponent is odd, you automatically know the answer will be either positive or negative i. Whatever the answer to the even-exponent part comes out to be determines whether the i will be positive or negative.

Although it is obvious that exponents which are multiples of 4 always yield the answer of "1", you cannot make this shortcut assumption for exponents of other multiples. For instance, all exponents which are multiples of 3 will not equal –i, as i^3 = -i, because the cycle repeats every forth number, not third. For proof, look at the solutions to i^6, i^{12} and i^{27}; their exponents are all multiples of 3, but they each have different results.

EVALUATING ROOTS WITH VARIOUS FACTORS & EXPONENTS

"Evaluate" is a fancy word for "Simplify as best you can, then give the answer."
There are a few steps here.

Remember how: When simplifying a radical (in which the radicand is only a number), you attempt to factor that number into "perfect-power" factors and whatever non-perfect power factor is left-over*? (This is possible according to the property of multiplying common bases with exponents). The same technique is used here, except now there may be multiple variable-factors to various powers as well. On that note, follow this **procedure**:

1. Break up all factors into perfect-power factors and non-perfect-power factors. This includes the coefficient as well as each variable.
2. Take the root of the perfect-power number; this comes out of the radical.
    *   Leave the non-perfect-power number under the radical.
3. For each variable raised to a perfect-power, take the root of it and move it outside the radical.
    *   An easy way to simplify multiples of higher powers is by dividing the perfect-power by the root. Move the base-variable raised to the quotient of the original power divided by the root, outside the radical.
    *   Leave the variables to non-perfect powers under the radical.

Use this procedure for the following **example**. Evaluate:
$$\sqrt{432x^6y^3z^4}$$

Since this is a square root (root 2), split each factor into perfect square factors, if possible, and whatever factors are left. It will look like:
$$\sqrt{(144)(3)x^6y^2y^1z^2}$$

Take the square roots of 144, x^6, y^2, and z^2, moving their simplified square roots outside as coefficients to the radical, and leave the remaining factors in the radicand:
$$12x^3yz\sqrt{3y^1}$$

Sometimes students aren't so sure how to take square roots of a base with an (even) exponent other than 2. In short, you are dividing the power of the base under the radical by the root-number. This is really thinking about it in *rational exponent* form. Take the isolated example of the square root of x^6: 6 divided by 2 (the root) equals 3. That's why x^3 is written outside the radical as a coefficient (among the others). Look at the square root of x^6 as a rational exponent:

$$x^{\frac{6}{2}}$$

The fraction reduces to 3, which, again, is why it is simplified to x^3.

Now evaluate this radical which is a cube root in the following **example**:

$$\sqrt[3]{135x^6y^3z^4}$$

Split each factor into perfect-cube factors, starting with the coefficient. It will look like:

$$\sqrt[3]{(27)(5)x^6y^3z^3z}$$

Take the cube roots of all the perfect cube factors and move them outside as coefficients to the radical, and leave the remaining non-perfect-cube factors in the radicand. Taking the cube roots of factors with exponents can be done by dividing their exponent by 3. The result simplifies to:

$$3x^2yz\sqrt[3]{5z}$$

MULTIPLYING BINOMIALS WITH RADICALS & i

Multiplying binomials with radicals or imaginary numbers is the same as multiplying binomials with variables. When new concepts involving radicals or imaginary numbers within complex numbers are introduced, sometimes students are intimidated because they view it as a completely new concept, when in fact it is hardly a new concept. If you can multiply binomials with variables, you can multiply binomials with radicals or imaginary numbers... by applying the FOIL method. You should consider the fact that imaginary numbers are just a subcategory of radicals. You just have to remember (to apply) the rule for multiplying radicals of the same root[*].

There is sometimes an additional step or two for multiplying binomials with radicals (including imaginary numbers). The product of certain radicals may be a radical which needs to be simplified. It may simplify to another term, still in radical form, or it may simplify to yield a perfect-root radicand, which will eliminate the radical. In that case, the last step will be to combine like-terms, which may be constants (numbers), or may contain variables, and because of this, the result may be a monomial, a binomial, a trinomial, or a polynomial with 4 terms, none of which are alike. Observe the similarities in the following **examples**, which will all follow the FOIL method.

Binomials with regular variables:

$(x + 3)(x - 5)$

$= x^2 + \text{-}5x + 3x - 15$

$= x^2 - 2x - 15$

Binomials with radicals that will have a simplified radical, but no like-radicals:

$\left(x + \sqrt{2y}\right)\left(x + \sqrt{4y}\right)$

$= x^2 + x\sqrt{4y} + x\sqrt{2y} + \sqrt{8y^2}$

$= x^2 + 2x\sqrt{y} + x\sqrt{2y} + 2y\sqrt{2}$

Binomials with variables and radicals where the middle radicals are alike, and the last radical simplifies to a constant:

$$\left(x + \sqrt{2}\right)\left(x + 3\sqrt{2}\right)$$

$$= x^2 + 3x\sqrt{2} + x\sqrt{2} + 3\sqrt{4}$$

$$= x^2 + 4x\sqrt{2} + 3(2)$$

$$= x^2 + 4x\sqrt{2} + 6$$

Binomials with radicals but no variables, where the middle radicals are alike, and the last radical simplifies to a constant, which is then combined with the left constant:

$$\left(5 + \sqrt{2}\right)\left(4 + \sqrt{2}\right)$$

$$= 25 + 5\sqrt{2} + 4\sqrt{2} + \sqrt{4}$$

$$= 25 + 9\sqrt{2} + 2$$

$$= 27 + 9\sqrt{2}$$

See more in Example 4 of Rationalizing the Denominator.

Binomials which are conjugate complex numbers, and result in just a constant, can be seen in (the denominator of) Example 5 of Rationalizing the Denominator.

RATIONALIZING THE DENOMINATOR

Rationalizing a denominator means manipulating a fraction so there is no radical in the denominator, while still maintaining the value of the fraction. Denominators are rationalized as a standard way to report fractions with radicals (this can include instances of "i" as well).

There are a few ways a radical may be in the denominator, as:
- as a monomial (1 term),
- in a binomial (2 terms, with one or two radicals),
  - it also may be as a complex number;

and the radical(s) may be a
- a square root, or
- a higher root

The method you will use to rationalize the denominator depends on how the radical is presented in the denominator.

Regardless of the type, you should recall these important concepts:
- *The definition of higher order roots*.
- Multiplying radicals*
- Conjugate binomials*
- Multiples of i, especially i^2 (which equals -1)
- Multiplying binomials with radicals, i, and complex numbers
- How to convert a fraction by multiplying the numerator & denominator by the same factors*

With that in mind, here is a general **procedure for rationalizing a denominator**:

You must recognize that there is a radical in the denominator. There only needs to be a minimum of one radical, but there may be more, and they may be of any root.

0. Simplify the radical[*], and the entire fraction, if possible. Although this is not absolutely necessary, it can allow you to work with smaller factors, and save you (from needing to recognize) an extra simplification step at the end.
1. What is the root-number? You will use this in step 3.
2. Also, what is the power? (It may be, and often is, an unwritten "1," but you still need to know this.) You will use this in the next step.
3. Determine: What power of the original radical (to be rationalized) must be multiplied times the radical, in order to eliminate the radical? (The definition of higher order roots gives the best clarity to this. Also, see the Examples and Note to follow for more details).
4. Is the denominator a monomial or a binomial?

4a. If it is a monomial, you only need to multiply it by another monomial.

4b. If the denominator is a binomial, you must multiply it by its conjugate binomial.

4c. Remember that a complex number (containing a term with "i") is considered a binomial with a radical, and will also be multiplied by its conjugate binomial.

- In this case, be prepared to get an intermediate-step term of i^2 which should be converted to "-1", which can then be combined with another constant, to complete the simplification.

Note: You don't need to be concerned with the numerator at all, except to do Step 5.

5. Be sure to multiply the numerator by the same factor you multiplied the denominator by.

There's also a chance that the denominator contains one radical but with multiple terms in the radicand. Let's assume the radical contains two terms, a binomial. In this case, you will proceed according to Step 3, except be prepared that when you multiply the radicands, you are just multiplying the denominator times (the proper number of multiples of) itself, not multiplying times a conjugate binomial.

For **example**, if the original rational expression to be simplified is:

$$\frac{1}{\sqrt{x+2}}$$

you will multiply the top and bottom by the square root of x *plus* 2:

$$\frac{1}{\sqrt{x+2}}\frac{\left(\sqrt{x+2}\right)}{\left(\sqrt{x+2}\right)} = \frac{\sqrt{x+2}}{x+2}$$

not the square root of the conjugate, x *minus* 2:

$$\frac{1}{\sqrt{x+2}}\frac{\left(\sqrt{x-2}\right)}{\left(\sqrt{x-2}\right)}$$ because this would yield $$\frac{\sqrt{x-2}}{\sqrt{x^2-4}}$$

which does not get rid of the radical in the denominator.

Note to Step 3: It may be extremely helpful to convert the radical to (or think of it in) rational exponent form, because, by looking at the root and power consolidated as one fractional exponent, you can answer the question from Step 3 by asking: What *fraction* must be added to this fraction so the sum of the fractions equals "1"? (because when multiplying bases with exponents, you add the exponents).

Explanation: If the radical is a *square root*, you must multiply it by *one factor* of itself (because a square root squared eliminates the radical and equals the radicand). If you think about it in rational exponent form, x to the one-half power must be multiplied by x to the one-half power, because one half plus one half equals one (whole).

Because square root radicals are so common, students often interpret this as "multiply the radical by *itself* to eliminate the radical," and mistakenly apply this technique to *all* radicals of various roots. But you must realize that this statement is only true by coincidence when applied to square roots. For instance:

- If the radical is a *cube root*, you must multiply it by *two factors* of itself (you may also think of it as "the cube root squared"). Multiplying this by the original cube root yields a cube root cubed, which correctly eliminates the radical and equals the radicand.

Observe the pattern in the following demonstrations:

- If the radical in the denominator is to root 2, multiply it by itself:

$$\frac{1}{\sqrt{x}}\frac{\left(\sqrt{x}\right)}{\left(\sqrt{x}\right)} = \frac{\sqrt{x}}{x}$$

- If the radical in the denominator is to root 3, multiply it by itself squared:

$$\frac{1}{\sqrt[3]{x}}\frac{\left(\sqrt{x}\right)}{\left(\sqrt{x}\right)}\frac{\left(\sqrt{x}\right)}{\left(\sqrt{x}\right)} \quad or \quad \frac{1}{\sqrt[3]{x}}\frac{\left(\sqrt[3]{x}\right)^{2}}{\left(\sqrt[3]{x}\right)^{2}} = \frac{\left(\sqrt[3]{x}\right)^{2}}{\left(\sqrt[3]{x}\right)^{3}} = \frac{\left(\sqrt[3]{x}\right)^{2}}{x}$$

- If the radical in the denominator is to root 4, multiply it by itself cubed.

$$\frac{1}{\sqrt[4]{x}}\frac{\left(\sqrt{x}\right)}{\left(\sqrt{x}\right)}\frac{\left(\sqrt{x}\right)}{\left(\sqrt{x}\right)}\frac{\left(\sqrt{x}\right)}{\left(\sqrt{x}\right)} \quad or \quad \frac{1}{\sqrt[4]{x}}\frac{\left(\sqrt[4]{x}\right)^{3}}{\left(\sqrt[4]{x}\right)^{3}} = \frac{\left(\sqrt[4]{x}\right)^{3}}{x}$$

- If the radical in the denominator is to root 5, multiply it by itself to the 4th power.

$$\frac{1}{\sqrt[5]{x}}\frac{\left(\sqrt{x}\right)}{\left(\sqrt{x}\right)}\frac{\left(\sqrt{x}\right)}{\left(\sqrt{x}\right)}\frac{\left(\sqrt{x}\right)}{\left(\sqrt{x}\right)}\frac{\left(\sqrt{x}\right)}{\left(\sqrt{x}\right)} \quad or \quad \frac{1}{\sqrt[5]{x}}\frac{\left(\sqrt[5]{x}\right)^{4}}{\left(\sqrt[5]{x}\right)^{4}} = \frac{\left(\sqrt[5]{x}\right)^{4}}{x}$$

- If the radical in the denominator is to root 6, multiply it by itself to the 5th power. Etc.

Example 1:

Simplify the following by rationalizing the denominator, which contains a *monomial square root*:

$$\frac{1}{\sqrt{3y}}$$

The denominator is of root 2 and is already simplified, so multiply the bottom and top by the square root of 3y:

$$\frac{1}{\sqrt{3y}}\frac{(\sqrt{3y})}{(\sqrt{3y})} = \frac{\sqrt{3y}}{3y}$$

Example 2:

Simplify the following by rationalizing the denominator, which contains a *monomial cube root*:

$$\frac{1}{\sqrt[3]{3y}}$$

The denominator is of root 3, so multiply the top and bottom by that denominator *squared*:

$$\frac{1}{\sqrt[3]{3y}}\frac{\left(\sqrt[3]{3y}\right)^2}{\left(\sqrt[3]{3y}\right)^2} = \frac{\left(\sqrt[3]{3y}\right)^2}{3y}$$

You might also think of it or write it as:

$$\frac{1}{\sqrt[3]{3y}} \cdot \frac{\sqrt[3]{3y}\,\sqrt[3]{3y}}{\sqrt[3]{3y}\,\sqrt[3]{3y}} = \frac{\left(\sqrt[3]{3y}\right)^2}{3y}$$

Example 3:

Simplify the following by rationalizing the denominator, which contains a *monomial cube root squared*:

$\dfrac{1}{\left(\sqrt[3]{3y}\right)^2}$, which is equivalent to $\dfrac{1}{\left(\sqrt[3]{9y^2}\right)}$ when the square is

in the radicand. The denominator is of root 3, but it is also already squared, so you only need to multiply the top and bottom by the missing factor (to make it a cube root cubed):

$$\dfrac{1}{\left(\sqrt[3]{3y}\right)^2} \cdot \dfrac{\sqrt[3]{3y}}{\sqrt[3]{3y}} = \dfrac{\sqrt[3]{3y}}{3y}$$

This is why Step 2 of the procedure reminds you to pay attention to the power as well as the root.

Example 4:

Simplify the following by rationalizing the denominator, which contains one *square root in a binomial*:

$$\dfrac{\sqrt{3y} + 5}{\sqrt{3y} - 5}$$

Put the binomials in parentheses. Multiply the top and bottom by the conjugate binomial of the denominator:

$$\dfrac{\left(\sqrt{3y} + 5\right)\left(\sqrt{3y} + 5\right)}{\left(\sqrt{3y} - 5\right)\left(\sqrt{3y} + 5\right)}$$

The makes the special case of the *product of conjugate binomials* in the bottom, but a *binomial squared* in the top:

$$\dfrac{\left(\sqrt{3y} + 5\right)\left(\sqrt{3y} + 5\right)}{\left(\sqrt{3y} - 5\right)\left(\sqrt{3y} + 5\right)} = \dfrac{3y + 10\sqrt{3y} + 25}{3y - 25}$$

Example 5:

Simplify the following by rationalizing the denominator, which contains a *complex number*:

$$\frac{1}{(5 + i)}$$

Multiply the top and bottom by the conjugate, which is "(5 - i)":

$$\frac{1}{(5 + i)}\frac{(5 - i)}{(5 - i)}$$

The top just becomes "5 – i"; FOIL the binomials in the bottom:

$$\frac{5 - i}{25 - 5i + 5i - i^2}$$

The top is simplified and will remain as-is until the end. In the bottom, the middle terms are combined and cancel each other out (as you would expect by multiplying conjugate binomials). Convert i^2 to negative one:

$$\frac{5 - i}{25 - (-1)}$$

The "minus negative one" becomes "plus one," leaving the simplified form:

$$\frac{5 - i}{26}$$

FINDING THE DOMAIN OF EVEN-ROOT RADICALS

Set *just* the contents of the radicand greater-than-or-equal to zero, as:

radicand ≥ 0,

and solve for x. Your answer will look something like:

$x \geq$ some #

The reason is: If the radicand of any even-root radical is any value *less than* zero, it makes the solution imaginary, and therefore wouldn't exist on a graph (at those x-values). Certain values of x which cause the radicand to be less than zero might also be considered *extraneous roots*[*].

This procedure does not apply to odd-root radicals because an odd root can be taken of a negative radicand.

SOLVING RADICAL EQUATIONS

Procedure for Solving a Radical Equation with *one radical*

1. Isolate the radical on the left, moving everything else to the right side using addition or subtraction.
2. Put parentheses around the *entire* right side (explained in Step 5b).
3. Square both sides (assuming that the radical is a square root. If the root is something other than 2, apply the proper power according to the definition of higher order roots).
4. On the left, it will remove the radical sign, leaving just the contents of the original radicand.

The right side may take an extra step to simplify, so be sure to bring down what you got in
Step 4 with each line.

5a. If the right side has one term, you will need to apply (distribute) the square to each factor, including the coefficient, if there is one. If any factors contain exponents, you will multiply each exponent times 2 (from the square).

5b. If the right side has two terms, this will become a *binomial squared*[*]. It is important that you treat it as such – You do not distribute the [2] and square each term (only), as many mistakenly do. In this case, you will have created a trinomial.

6. Simplify. Combine like terms.
   - If the highest degree term is just x (to the unwritten power of 1), isolate it, move the constant to the other side... you will have your answer, and you are done. However...

7. If the highest degree term is degree 2 (as x^2), move all terms to one side, arranged in standard form & descending order.

8. Check to see if the polynomial can be factored, perhaps into two binomials.
   - If so, factor, and solve each group.
   - If not, use the Quadratic Formula to solve.

9. Check your answers by substituting each one back into the original equation, and simplifying. If either one comes out to be where one side does not equal the other, you have found an extraneous root, and you must eliminate this answer. Extraneous roots are common about 20% of the time.

You can see examples of this midway through examples involving multiple radicals (that reduce to having one radical) in:

- the Annotated Example involving *three radicals*, each of which *contain binomials*, and
- the Annotated Example involving *two radicals*, ending as a *quadratic*.

Procedure for Solving a Radical Equation w/ *two or three radicals*

This is the **procedure** for a radical equation in which one radical on the left is equal to a binomial on the right, which either contains one or two radicals.

1. Isolate one of the radicals on the left, moving everything else to the right side.

2. Put parentheses around the *entire* right side (explained in Step 5).

3. Square both sides, assuming that the radicals are square-roots. If the root is something other than 2, apply the proper power according to the definition of higher order roots.

On the left, it will remove the radical sign, leaving just the contents of the original radicand. The right side may take an extra step to simplify, so be sure to bring down what you got on the left from Step 3, with each line.

4. The right side will become a *binomial squared*[*]. It is important that you treat it as such – You do not distribute the ^{2} and square each term (only), as many mistakenly do. In this case, you will have created a trinomial with *one radical*.

5. Proceed by following the Procedure for Solving a Radical Equation with *one radical,* to completion.

Note: If you are given an equation with four radicals, you will start by putting two radicals on each side, giving you a binomial on each side. Put both sides in parentheses and square both sides (both binomials). Each side will then result in a trinomial with a radical in the middle term. If the radicals from both sides are like-terms, combine them, then isolate the consolidated radical, and take the square root of both sides. If the two radicals are not alike and cannot be combined, isolate one of the radicals, and proceed from Step 3.

Annotated Example involving *three radicals*, each of which *contain binomials*

$$\sqrt{4x - 11} - \sqrt{x - 1} = \sqrt{x - 4}$$

As it is given, one radical, the "$\sqrt{x - 4}$" is already isolated. For the purpose of consistency, let's add "$\sqrt{x - 1}$" to both sides, to so the isolated radical is on the left.

$$\sqrt{4x - 11} - \sqrt{x - 1} + \sqrt{x - 1} = \sqrt{x - 4} + \sqrt{x - 1}$$

$$\sqrt{4x - 11} = \sqrt{x - 1} + \sqrt{x - 4}$$

Put parentheses around both sides:
$$\left(\sqrt{4x - 11}\right) = \left(\sqrt{x - 1} + \sqrt{x - 4}\right)$$

Square both sides.
$$\left(\sqrt{4x - 11}\right)^2 = \left(\sqrt{x - 1} + \sqrt{x - 4}\right)^2$$

Notice that when the right side is squared, it makes a *binomial squared*, as there are two radicals, which are the two terms.
On the left, the square root radical is removed, according to the definition of higher order roots.

$$4x - 11 =$$

Since the right is a binomial squared, we will follow the special case procedure:

Square the first term: $\left(\sqrt{x - 1}\right)^2$ which becomes $x - 1$

Multiply the first term times the last term, times 2:
$$2\left(\sqrt{x - 1}\right)\left(\sqrt{x - 4}\right)$$

- This follows the product rule of radicals, which states that when you multiply radicals with the same root, you can multiply their radicands under the same root.
$$2\left(\sqrt{(x - 1)(x - 4)}\right)$$

- Now, under the radical, you are multiplying two binomials. You can FOIL these.
  - Product of the First terms is x^2
  - Product of the Outer terms is -4x
  - Product of the Inner terms is –x
  - Product of the Last terms is +4
  - The sum of the products of the Outer & Inner terms -4x – x is -5x, the middle term
  - These stay under the radical as: $\sqrt{x^2 - 5x + 4}$

Square the last term: $\left(\sqrt{x - 4}\right)^2$ which becomes x - 4

Putting these all in the equation looks like:
$$4x - 11 = x - 1 + 2\left(\sqrt{x^2 - 5x + 4}\right) + x - 4$$

Simplify. Combine like-terms on the right:
- x and x combine as 2x
- -1 – 4 combine as -5

It now looks like:
$$4x - 11 = 2x - 5 + 2\left(\sqrt{x^2 - 5x + 4}\right)$$

Notice that the right side required more steps to simplify and the "4x – 11" was simply carried down with each step.

Continue simplifying, by continuing to combine like-terms from both sides. Let's move all terms, except the one with the radical, to the left. Subtract 2x and add 5 to both sides, moving them to the left.
$$4x - 11 - \mathbf{2x} + \mathbf{5} = 2x - 5 + 2\left(\sqrt{x^2 - 5x + 4}\right) - \mathbf{2x} + \mathbf{5}$$

(On the left, 4x – 2x becomes 2x, and -11 + 5 becomes -6)

It now becomes:

$$2x - 6 = 2\left(\sqrt{x^2 - 5x + 4}\right)$$

This is continued on the next page…

The term with the radical is isolated, but now we want to isolate just the radical. We will do this by dividing both sides by the coefficient 2:

$$\frac{2x - 6}{2} = \frac{2(\sqrt{x^2 - 5x + 4})}{2}$$

On the left, factor out the GCF of 2 in the numerator:

$$\frac{2(x - 3)}{2} = \frac{2(\sqrt{x^2 - 5x + 4})}{2}$$

which allows you to cancel out the 2 in the top and bottom, on both the left and the right:

$$x - 3 = \sqrt{x^2 - 5x + 4}$$

Put both sides in parentheses to set up for the next step:

$$(x - 3) = \left(\sqrt{x^2 - 5x + 4}\right)$$

Square both sides:

$$(x - 3)^2 = \left(\sqrt{x^2 - 5x + 4}\right)^2$$

making the left side a binomial squared, and removing the radical on the right:

$$x^2 - 6x + 9 = x^2 - 5x + 4$$

Simplify: Combine like-terms. You would normally move all terms to the same side, and set equal to zero, in case it set up a quadratic… but since both x^2 terms cancel each other out, you can put the terms with x on the left and the constant(s) on the right, as we would have to position them this way anyway to solve for x.

$$x^2 - 6x + 9 \ \mathbf{-x^2 + 5x - 9} = x^2 - 5x + 4 \ \mathbf{-x^2 + 5x - 9}$$

In this case, we added 5x to both sides (to move it to the left), and subtracted 9 from both sides, moving it to the right. Combine like-terms. It simplifies to:

$$-x = -5$$

Divide both sides by -1 to isolate x, and

$$x = 5$$

This appears to be the answer, but we're not quite finished…

Check your answer by substituting 5 back in for each x in the original equation:

$$\sqrt{4(5) - 11} - \sqrt{5 - 1} = \sqrt{5 - 4}$$

$$\sqrt{20 - 11} - \sqrt{4} = \sqrt{1}$$

$$\sqrt{9} - \sqrt{4} = \sqrt{1}$$

$$3 - 2 = 1$$

$$1 = 1$$

The left equals the right, meaning it checks out. Therefore, we accept x = 5 as the answer.

Note: You can see that this example is somewhat long and tedious. You may not need to write out every single step, as I did, but I showed every detail to prevent any instance of confusion. You might also notice that a number of steps were repeated where necessary. Be sure to leave yourself plenty of room, go step-by-step, and don't get frustrated by the lengthiness of the problem. Remember to check your answer.

Annotated Example involving *two radicals*, ending as a *quadratic*

$\sqrt{2x + 5} - 2\sqrt{2x} = 1$

Add the term $2\sqrt{2x}$ to both sides to move it to the other side,

$\sqrt{2x + 5} - 2\sqrt{2x} + 2\sqrt{2x} = 1 + 2\sqrt{2x}$

which isolates one radical on the left. Put each entire side in parentheses:
$\left(\sqrt{2x + 5}\right) = \left(1 + 2\sqrt{2}\right)$

Square both sides:
$\left(\sqrt{2x + 5}\right)^2 = \left(1 + 2\sqrt{2}\right)^2$

The left side becomes:
$2x + 5 =$

as the radical is removed. The right side becomes a binomial squared:
$2x + 5 = 1 + 4\sqrt{2x} + 4(2x)$

creating a radical in the middle term.

Simplify: The 4(2x) becomes 8x. Move all terms, except the radical, to the left by subtracting 1 and 8x from both sides:
$2x + 5 - 1 - 8x = 4\sqrt{2x} + 1 + 8x - 1 - 8x$

Combine like terms, and the equation becomes:
$-6x + 4 = 4\sqrt{2x}$
Divide both sides by the coefficient 4; on the right, the 4s cancel out which isolates the radical:
$\dfrac{-6x + 4}{4} = \dfrac{4\sqrt{2x}}{4}$

Intermediate step to simplify: On the right, factor the GCF of 2 out of the numerator:

$\dfrac{2(-3x + 2)}{4} = \sqrt{2x}$

On the left, factor "2" out of the top and bottom, which becomes:

$$\frac{-3x + 2}{2} = \sqrt{2x}$$

Put both sides in parentheses and square both sides:

$$\left(\frac{-3x + 2}{2}\right)^2 = \left(\sqrt{2x}\right)^2$$

On the left, the power of 2 gets distributed to the binomial on top and the 2 on the bottom.
The top of the left becomes a binomial squared. The bottom 2 is squared to become 4.
The radical on the right is removed. The equation becomes:

$$\frac{9x^2 - 12x + 4}{4} = 2x$$

(Think of the 2x on the right as being over 1). Cross multiply to eliminate the fraction on the left. The denominator 4 multiplied by the 2x on the right becomes 8x on the right; the trinomial from the numerator on the left is multiplied by 1 (from under the 2x), which remains the same trinomial on the left.

$$9x^2 - 12x + 4 = 8x$$

Simplify in order to put the quadratic equation in standard form.
Subtract 8x from both sides:
$$9x^2 - 12x + 4 - \mathbf{8x} = 8x - \mathbf{8x}$$

You now have the quadratic equation:
$$9x^2 - 20x + 4 = 0$$

Attempt to factor and solve; if it can't be factored, put it in the quadratic formula… but it can be factored into:
$$(9x - 2)(x - 2) = 0$$

Set each binomial equal to zero and solve for x.

For the left binomial factor:
9x – 2 = 0
9x – 2 + **2** = 0 + **2**
9x = 2, divide both sides by 2.

The apparent solutions are:

$x = \frac{2}{9}$, and $x = 2$, however, these solutions *must* be checked by substituting each solution into the original equation and simplifying. First, test $\frac{2}{9}$:

$$\sqrt{\left(2\left(\frac{2}{9}\right) + 5\right)} - 2\sqrt{2\left(\frac{2}{9}\right)} = 1$$

$$\sqrt{\left(\frac{4}{9} + 5\right)} - 2\sqrt{\frac{4}{9}} = 1$$

Convert the 5 into a like-fraction with 9 in its denominator:

$$\sqrt{\left(\frac{4}{9} + \frac{5}{1} \cdot \frac{9}{9}\right)} - 2\sqrt{\frac{4}{9}} = 1$$

Now the two like-fractions in the left radical can be added…

$$\sqrt{\left(\frac{4}{9} + \frac{45}{9}\right)} - 2\sqrt{\frac{4}{9}} = 1$$

The left radical has one fraction from the addition of 4 and 45 becoming 49. Both radicands are now perfect squares. Remember to distribute (apply) the radical to the numerators and denominators.

$$\sqrt{\frac{49}{9}} - 2\sqrt{\frac{4}{9}} = 1$$

The square roots are taken and the radicals are removed. The rest is just simplification…

$$\frac{7}{3} - 2\left(\frac{2}{3}\right) = 1$$

Subtract the like fractions, as they both have the common denominator of 3:

$$\frac{7}{3} - \frac{4}{3} = 1$$

which becomes:

$$\frac{3}{3} = 1$$

and reduces to:

$1 = 1$, therefore $\frac{2}{9}$ is an acceptable solution.

Now check x = 2 by substituting it into each x in the original, and simplifying:

$$\sqrt{2(2) + 5} - 2\sqrt{2(2)} = 1$$

$$\sqrt{4 + 5} - 2\sqrt{4} = 1$$

$$\sqrt{9} - 2(2) = 1$$

$$3 - 4 = 1$$

$-1 \neq 1$, therefore 2 *is not* a solution.

The only solution is $\frac{2}{9}$.

Multiplying Two Radicals with Binomial Radicands

When multiplying two radicals (of the same root) with binomials in the radicand, you will have an extra step to do than if the radicands were monomials (single terms). You must follow the *product rule for radicals*, which will then lead you to *multiplying two binomials* (in which case you could use the FOIL method) under the radical.

This is shown in the middle of the previous example.

CONSTANTS & COEFFICIENTS DON'T MIX

Most books (including this one) spend the majority of the pages showing what *to do*. Being an instructor, you get to know common mistakes that students make, for no particular, predictable reason, other than: students tend to make the same mistakes over and over. A unique feature of *ALGEBRA IN WORDS* is to draw attention to those mistakes in order to prevent them early on.

Sometimes (with anything, not just algebra), if you get accustomed to doing something the incorrect way in the beginning, it has a way of getting ingrained in your memory, and the longer you do it, the harder it is to undo and correct.

I want to draw your attention to a mistake students often make involving the addition or subtraction of a constant and a radical with a coefficient. In short, constants and radicals (and their coefficients) are not like-terms[*], and therefore cannot be combined.

Take the following example:
$3 + 2\sqrt{5x}$

This is already in simplified form.
It is *incorrect* to add the constant 3 and the coefficient 2 to get
$5 + \sqrt{5x}$ *or* $5\sqrt{5x}$

Do not do that. Constants and coefficients don't mix because they are simply not "like-terms".
This might occur in any step of a problem involving radicals. It will likely be seen in:
- Complex numbers
- Fractions in which you must rationalize the denominator
- Completing the Square
- The end steps of using the Quadratic Formula[*]
- The Pythagorean Theorem
- Radical Equations

Be cautious not to commit this mistake. It is easy to avoid, especially when it's already simplified.

COMPLETING THE SQUARE

Completing the Square is one method used to solve quadratic equations*. It is also used to make the Standard Form Equation of a Vertical Parabola and the Standard Form Equation of a Circle.

This method of solving a quadratic equation is sometimes chosen once a quadratic equation is arranged in standard form, and it is determined that it can't be factored, however the Complete the Square Method may be used even if the quadratic can be factored. The goal is to create a *perfect square trinomial* on the left, set equal to some number on the right. The perfect square trinomial on the left is created specifically to be factored into a *binomial squared*. Then, the square root of both sides is taken. One additional step is taken to result in two solutions, which may be "real" or "complex number" solutions.

Procedure:
0a. Arrange the equation into standard form: $x^2 + bx + c = 0$
0b. Be sure the coefficient of the leading term is positive 1. If it is not, divide each term by the current coefficient (a).

1. Move the c-term to the other side of the equation (by properly adding or subtracting its opposite to both sides). Make a space to the right of bx and to the right of –c (this is explained a few steps later). Now your equation will be in the form:

    $$x^2 + bx + \underline{\quad} = -c + \underline{\quad}$$

Note: The value of c is not necessarily negative, but there is a "-" in front of c, here, to show it was moved from the left side to the right side. Also, there is no coefficient "a" because it should now be "1", and we usually don't write coefficients of "1".

2. You will now find the "new c" value by using the following formula:
    $$\left(\frac{b}{2}\right)^2 = \text{new c}$$

(where b is the coefficient in front of x from the bx term).
*   First, divide the value of b by 2, then square that number; that is your "new c" value).

3. Add the "new c" *to both sides*:

$$x^2 + bx + new\ c = -c + new\ c$$

4. Factor the left side into a binomial squared, and combine the c-numbers on the right, which will become a new number:

$$\left(x + \frac{b}{2}\right)^2 = \#$$

For ease of following this procedure, I'm going to replace $\frac{b}{2}$ with the square root of the "new c" since they will be the same number:

$$\left(x + \sqrt{new\ c}\right)^2 = \#$$

5. Take the square root of both sides. Remember that (on the right) the square root of a number yields a positive and negative result.

$$\sqrt{\left(x + \sqrt{new\ c}\right)^2} = \sqrt{\#}$$

The left side equals the contents of the parentheses, based on the definition of higher order roots:

$$x + \sqrt{new\ c} = \pm\sqrt{\#}$$

6. Subtract $\sqrt{new\ c}$ from both sides, moving it to the right,

$$x + \sqrt{new\ c} - \sqrt{new\ c} = \pm\sqrt{\#} - \sqrt{new\ c}$$

thus isolating x:

$$x = -\sqrt{new\ c} \pm \sqrt{\#}$$

The "$-\sqrt{new\ c}$" is written first, on the right, so that you can *add and subtract* $\sqrt{\#}$ from the number you just moved to the right. The "plus-or-minus" segment is usually written at the end of the statement. Also, the sign in front of the $\sqrt{new\ c}$ may be positive or negative, depending on the sign in front of the original "b".

7. You should now simplify $\sqrt{\#}$ if you can. Nearing the conclusion, you may get three types of solutions:

7a. If the # inside $\sqrt{\#}$ is a positive perfect square (1, 4, 9, 16, etc.), take the square root of it. Since this yields an integer (a non-decimal number), you should break up the equation for the "+" route and the "-" route, as:

$$x = -\sqrt{new\ c} + \sqrt{\#}\ , \text{and}$$
$$x = -\sqrt{new\ c} - \sqrt{\#}$$

This will result in two real, non-decimal numeric solutions.

7b. If the # in your $\sqrt{\#}$ is a positive non-perfect square (like 2, 3, 5, 12, 20, etc), simplify it, if possible. Regardless, you will have a radical, which is usually left in the radical form. In this case, it is acceptable to report both solutions in the form:

$$x = \#\ \pm\ \#\sqrt{\#}$$

(In my representation, the "#s" may all be different non-zero numbers, and may be 1).
This results in two real, but irrational, solutions.

Also, if you are using this method to find and graph x-intercepts, you should take the square root of the number on your calculator, resulting in an approximated (rounded) decimal, which should then be *added and subtracted* to the first number, yielding **two real, approximated solutions in decimal form**.

7c. If the # in your $\sqrt{\#}$ is negative, you must factor out the $\sqrt{-1}$ and replace it with "i", and simplify the remaining factor. Since part of the solution contains an imaginary number, this results in **two conjugate, complex number solutions**.

Using this knowledge, you should be able to recognize a perfect square trinomial…

RECOGNIZING A PERFECT SQUARE TRINOMIAL

Once a trinomial is in standard form:

$ax^2 + bx + c = 0$...

Analyze b with respect to a and c...

Ask yourself: Is b the product of:

$\sqrt{a}$ times $\sqrt{c}$ times 2 ?

If *it is*, that is enough criteria to say that it is a Perfect Square Trinomial, and therefore can be factored into a binomial squared. Often, the a and c numbers will be obvious perfect-squares (like 1, 4, 9, 16, etc.), as well, and you might notice them first before analyzing b.

WHERE DID THE QUADRATIC FORMULA COME FROM?

If you are ever asked where the quadratic formula comes from, or if you are ever asked to derive it, the Quadratic Formula comes from the generic standard form of a quadratic equation:

$ax^2 + bx + c = 0$,

then solved for x by the Complete the Square Method.

In the intermediate steps, a *binomial squared* is made on the left, and then the *square root property* is applied (taking the square root of both sides).

It is very likely that your textbook shows the entire "proof" or "derivation."

TRANSFORMING A NON-QUADRATIC into a QUADRATIC EQUATION using SUBSTITUTION, then SOLVING

These types of equations are often given with three terms, where the exponent of the leading term is either a multiple of 2 (could be a negative multiple), or a rational exponent (fraction) in relation to 2, and the exponent of the middle term is half of the exponent of the leading term.

You will be given equations like this where you are expected to solve them. The trick is that they are likely related to quadratic equations by the pattern of their exponents. In short, you must convert it into a quadratic equation using substitution (because it is fairly easy to solve quadratics*), solve it (for "m"), convert it back to the original (pre-substitution) equation, and solve the rest. Remember that, like any equation, you can expect to have a maximum number of solutions a the degree of the equation (unless the exponents are fractions, in which case you will probably get two solutions, if they are both valid).

Follow this more detailed procedure to properly convert a non-quadratic equation into a quadratic, and solve for all possible solutions. Note: Many textbooks have you use "y" as the substituted variable, however, I recommend you use "m", to prevent confusion, since y is often used in other contexts.

Procedure:

0. Evaluate and ensure that the original equation is in descending order and standard form (unless the exponents are fractions or negative, in which case the first two terms should be arranged in ascending order, with the constant always last). If there is a GCF, factor it out, and apply the following instructions to the remaining trinomial.

1. Let m = "x to the power of the middle term". Write this step out, so you know what to substitute back in later.

2. Substitute "m^2" in for the x-portion of the leading, left-most term; keep the original coefficient.

3. Substitute "m" in for the x-portion of the middle term, and keep its original coefficient.

4. Re-write the constant (the last number), as it does not change.

You should now have a quadratic equation in the form of:
$$am^2 + bm + c = 0$$

5. Solve for m, either by Factoring or by the Quadratic Formula. This may result in two solutions.

6. Substitute x and its original exponent back in for m in each of the solutions from the previous step.

7. Solve for x in each instance. You may need to solve terms with rational exponents.

8. Evaluate and simplify completely:

8a. Is there a GCF in the numerator? If so, factor it out.

8b. Is there a common factor in the numerator and denominator that can be cancelled out? If so, cancel them out.

8c. Is there a radical in the denominator? If so, it must be rationalized.

Check your answers. You could do this after Step 7, before Step 8; it might save you some work if you find an extraneous solution.

FACTORING TRINOMIALS USING THE ac/GROUPING METHOD

This method is often used when the numbers in a trinomial are too large for you to factor using the Trial & Error Method. This method may take up a lot of paper space, so leave yourself ample room before beginning.

Procedure:
0. Arrange your terms into descending order and standard form, if they are not already. As the heading suggests, there are two parts to this: The ac portion and the grouping portion.

The "ac" portion of the procedure:

1. Multiply the coefficient "a" times the constant "c". This product is "ac".

2. Somewhere on your paper, make a table with two columns looking like:

ac factors	Sum of factors

It will have as many rows as there are possible factors.

2a. In the left column, write out all the possible integer factors (including their signs) of the product of ac.

2b. In the right column, write the sum of those factors.

3. Your goal is to scan through the right column to find:
Which set of factors, which, when added, yield "b" from the original trinomial? You will use these factors in the next step.

4. Write the first (ax^2) and last (c) terms of the original trinomial, leaving space for two terms between them, and leaving out the "bx" term, like:

$$\#x^2 \underline{\hspace{2cm}} \underline{\hspace{2cm}} + \#$$

5. Take the factors you determined from Step 3, attach "x" to each factor, and strategically write them in the empty spaces above.

- Of the two x-terms, write the term that has a common factor with the c-term in the right space. This is essential to Step 7. (Notice that either x-term can be written in the left space, as it already has a common factor with ax^2, that being "x".

This should now look something like:

$\#x^2 + \#x + \#x + \#$ (where each "#" may be different numbers)

The Grouping portion of the procedure:

6. Follow this instruction very carefully:
Put the two left terms in parentheses,
then write a "+" sign,
then put the two right terms in parentheses, being sure to keep the sign of the x-term *inside* the parentheses. (This is a step where many mistakenly put the right set of parentheses to the right of the sign between the two x-terms. This mistake is explained in FMMs.)
It should now look like:

$(ax^2 + \#x) + (+\#x + c)$

7. Look for a GCF in each set of parentheses, separately (the GCFs will likely be different in each group).

8. Factor the GCF out in each group, respectively. (Note, if you can't find a GCF in one or both groups, switch the position of x-terms from Step 5, then re-check). If this is done correctly, you should notice that the parenthetical groups are the same; *this* is the *new (third) GCF*, which will be *factored out* in the next step.

9. Set up two sets of parentheses next to each other, as:
( )( )

The left group of parentheses should be the (third) common-factor-group; the right group will contain each of the GCFs from each of the two parenthetical groups from Step 7. The trinomial is now factored.

10. If the original trinomial is part of an equation (set equal to 0), you must solve. Set the contents of each set of parentheses equal to zero and solve for x.

PROCEDURE FOR GRAPHING A QUADRATIC EQUATION (PARABOLA)

0. Make sure the equation is arranged in standard form and descending order.

1. Find the vertex:
   - First by finding x, using $x = \frac{-b}{2a}$ from $ax^2 + bx + c$
   - Then use x to find its associated y by substituting the value of x into the original equation [f(x)] and solving.
   - Write the x and y in parentheses as an ordered pair.

2. Identify the y-intercept; it is "c" from $(ax^2 + bx + c)$.

3. Solve for the x-intercepts by either factoring and solving, or by using the Quadratic Formula. If your solutions contain radicals, you should compute them to get approximated, decimal numbers. This will give you more manageable numbers to graph.

4. As an extra clue, the sign of "a" indicates whether the parabola is positive (opens upward), or negative (opens downward).

5. If there are no real solutions found in Step 3, you must find additional points to generate an accurate sketch. A good place to start is by looking at the vertex and using values in single units above or below the x-value of the vertex (you don't need to use x = 0 because you already have it - that's the y-intercept). For instance, if the x-value of your vertex is "3" then you may choose to find y when x is 1, 2, 4 and 5.

6. Sketch the parabola by connecting the points you have found with smooth, curved lines. Don't connect them with sharp, jagged lines, making it look like a V. Also, clearly show the curve going through all the points you have found, as your instructor will likely look for them when your work is evaluated. You might even label your points.

STANDARD FORM EQUATION OF A VERTICAL PARABOLA

Thus far, you have probably gotten used to the standard form for a quadratic equation. Sometimes you will have to convert equations for parabolas into standard form. The standard form equation of a vertical parabola is:

$y = a(x - h)^2 + k$

where:
a is the coefficient, (which cannot be 0) to show shrinking or stretching, and whose sign tells if the parabola opens upward or downward;
h is the x-value representing the number of horizontal units the vertex is from the origin; and
k is the y-value representing the number of vertical units the vertex is from the origin.
Together, (h, k) is the vertex of the parabola.
Also, h is the (vertical) axis of symmetry (by the equation "x = h", which, if drawn, is drawn as a dotted line).

When the vertex is at the origin, (and there is no coefficient for shrinking or stretching), the standard form equation could be written as:

$y = a(x - 0)^2 + 0$

or

$y = ax^2$

Note: h & k are used here, but they are also used in different contexts for other types of graphs and equations. Generically, h represents the horizontal shift from the origin and k represents the vertical shift from the origin.

The following is the **procedure** for converting a quadratic equation in standard form into the standard form equation for a vertical parabola. In short, you will be using the Complete the Square Method to create a trinomial, which will then be factored into a binomial squared. This is why you must apply the first step:

1. To start, the coefficient in front of x^2 must be 1. If it is (an unwritten) 1 to start, proceed to Step 2. If it is anything other than 1, put parentheses around ax^2 and bx, together, but leave c out of the parentheses, then factor "a" out of the left group. This may turn the coefficient b into a fraction, which is okay. It will look like:

$$y = a\left(x^2 + \frac{b}{a}x\right) + c$$

2. Inside parentheses, you will perform the Complete the Square Method, but in a slightly different way. (Note: For simplicity, I'm going to refer to "b-over-a" as just "b".) In short, you will be adding *and* subtracting the *new c* on the *same side* of the = sign. You will add the new c inside the parentheses and subtract the new c outside the parentheses. Leave a distinct space for each new c so you don't forget to write it. Set it up like this:

$y = a(x^2 + bx + \underline{\quad}) - \underline{\quad} + c$

Fill it in with the proper numbers by finding the *new c*.

$y = a(x^2 + bx + \text{new c}) - \text{new c} + c$

3. Inside the parentheses, you have created a perfect square trinomial which should now be factored into a binomial squared. The term on the right in the parentheses is now the number "h".

4. On the outside, on the right, combine the constants. The subtracted new c combined with the original c now make the number "k". It should now look like:

$y = \#(x - \#)^2 + \#$

Note 1: The sign in front of h may be positive. If it is, that signifies that the value of h is negative (because subtracting a negative becomes adding). Likewise, the sign in front of k may be negative. If it is, that signifies that the value of k is negative (because adding a negative is like subtracting).

Note 2: You may be used to using c from the standard form of a quadratic equation for the y-intercept. However, in the standard form equation for a vertical parabola, even though "k" is a stand-alone constant, it is not the y-intercept. When the equation is in standard form for a (vertical) parabola, and you want the y-intercept, you can either substitute 0 in for x and solve for y, or convert the equation into standard form and use c.

Note 3: This procedure applies to vertical parabolas. The procedure for converting the equation of a horizontal parabola into standard form is not covered in this book.

QUADRATIC INEQUALITIES

You will be given a quadratic equation that is either $>$, $\geq$, $<$, or ≤ 0. You should first understand what this means. First, think of "0" as the x-axis (from the equation $y = 0$)[*] because...
When the inequality is set to *greater-than* 0, it's like asking: What are all the x-values *above* the x-axis? And, when the inequality is set to *less-than-or-equal-to* 0, it's like asking: What are all the x-values *on or below* the x-axis? Etc.

The solution(s) to a *quadratic inequality* are the domain(s) or region(s) of x that make the inequality statement true... In other words: all possible values for x that satisfy the constraints of an inequality statement.

The solution(s) are reported either in *interval notation* or *set-builder notation*, and you may be required to support your answer with a sketch on a number-line.

An *interval* is a defined region. Intervals are separated by *boundary points*. Boundary points are the intermediate "solutions" you will find in Step 2 of the procedure (you also might know them as x-intercepts on a 2D graph). "Testing an interval" means testing to see if parts of the graph are above or below the x-axis on a 2D graph.

Note: The examples in most textbooks have you view only the x-axis by drawing the number-line (which, in essence, is the x-axis), but you will gain a clearer understanding of what's going on (and get the same conclusions) by drawing the actual (2D) graph, as I advise you to do later in the procedure section.

Procedure (on a 1D Graph)

0. Make sure the inequality is simplified and in standard form.

1. Change the inequality to an equation by switching the inequality symbol with an "=" sign.
    This should now be a quadratic equation.

2. Solve the quadratic equation either by factoring (if possible) or by the Quadratic Formula.
    *   The solutions to the quadratic equation are the *boundary points* (and are used in the next steps, but these are not the solutions to the problem, so you must continue on).

3. Draw a number line. This is a 1D Graph representing the x-axis.

4. Plot your x-intercepts (now known as boundary points) on the number-line; be sure to use the proper inclusive/noninclusive symbols (open/closed-dot, or parentheses/brackets). Any area to the left, right, or between boundary points are considered regions or intervals.

5. Below each interval/region, define each region in reference to x. Write them out in interval notation or as inequalities. In the next steps, you will test each region, which will either Pass or Fail the tests.

6. Pick an obvious test point (an x-value) in each region; these test points cannot be the same numbers as the x-intercepts (boundary points) you found in Step 2 and plotted in Step 4.

7. Substitute each test point into the original inequality (or the one you simplified and rearranged), then simplify…

8. If the test (simplified inequality statement) comes out True, that indicates that the region/interval represented by that test point is a valid (part of the) solution;
    and if the test comes out False, then you reject that region/interval as part of the solution. Your tests will likely reveal regions that both pass and fail, but see Notes 1 & 2.

Continued with Step 9 on the next page…

9. Based on your conclusions, present your answer in interval notation, set-builder notation, and/or graphical (number-line) form. Be sure to properly use "and" or "or" in your solution, either by applying the proper constraints or assumptions for "and" & "or".

Note 1: If there is only one x-intercept, this means that the parabola is touching the x-axis. In this case, the solutions may either be:
- all real numbers,
- all real numbers except for the x-intercept,
- no solution, or
- just the x-intercept.

The solution will depend on:
- whether the parabola opens upward or downward,
- the inequality symbol, and
- the inclusive/noninclusiveness of the inequality symbol.

Note 2: If there are no x-intercepts, the solution may either be:
- all real numbers, or
- no solution.

You will know the solution can only be one of these when you only find imaginary number solutions when you solve for the x-intercepts. The solution will depend on:
- whether the parabola opens upward or downward, and
- the inequality symbol.

The inclusive/noninclusiveness won't affect the solution in this case.

Get a clearer understanding of this by sketching and looking at the 2D Graph perspective...

Procedure (on a 2D Graph)

After performing steps 0 through 2 in the Procedure on a 1D Graph...

3. Plot the x-intercepts on a full 2 dimensional graph.

4. Sketch the parabola, using the x-intercepts, the y-intercept (c), and the sign of "a" to determine if it opens upward or downward. The x-intercepts still act as boundary points.

5. Define each region/interval in reference to x.

6a. If the inequality is > or ≥ to 0, shade the region(s) in the graph *above* the x-axis.
6b. If the inequality is < or ≤ 0, shade the region(s) in the graph *below* the x-axis.

The region with the shaded part is the solution to the problem. Report the domain of the shaded region in interval or set-builder notation. This method bypasses the need to use the constraints or assumptions for "and" & "or" because you can just see the solution on the graph.

Example 1 using the 1D procedure:
$x^2 + 4x - 21 > 0$

All terms are on the left with zero on the right, as we want them.
Convert this inequality to an equation:
$x^2 + 4x - 21 = 0$

Find the x-intercepts (boundary points). Attempt to solve, first, by factoring; it can be factored into:
$(x - 3)(x + 7) = 0$

The x-intercepts (boundary points) are:
$x = 3$ and $x = -7$

Plot these on a number-line /1D graph (not shown here). Since the inequality is noninclusive, use open-dots or parentheses. Now define each of the 3 intervals (in interval notation):
$(-\infty, -7), (-7, 3), (3, \infty)$

Choose test points for each region that are not on the boundary points:
Let's use: -8, 0, and 4

Substitute them into the original inequality to test the truth of the statement:
-8: $(-8)^2 + 4(8) - 21 > 0$
$16 + 32 - 21 > 0$
$27 > 0$ is true, therefore $(-\infty, -7)$ *is* a solution.

0: $0^2 + 4(0) - 21 > 0$
$-21 > 0$ is untrue, therefore $(-7, 3)$ is *not* a solution.

4: $4^2 + 4(4) - 21 > 0$
$16 + 16 - 21 > 0$
$11 > 0$ is true, therefore $(3, \infty)$ *is* a solution.

If you approach the solution from a constraints perspective, here's how you know to answer with "or"... substituting either a number within $(-\infty, -7)$ *or* $(3, \infty)$ makes the inequality statement true. These were already tested using test points "-8" and "4" in previous steps. Also, since the inequality in the original statement is ">", and the parabola is positive, you can assume to combine the interval solutions with "or". The solution, in interval notation, is: $(-\infty, -7) \cup (3, \infty)$
In set-builder, consolidated form: $\{x|\ -7 < x > 3\}$

108

Example 2 using the 1D procedure:

$$x^2 + 2x \leq 5$$

Move all terms to one side (the left) by subtracting 5 from both sides, to get:

$$x^2 + 2x - 5 \leq 0$$

Convert this to an equation:

$$x^2 + 2x - 5 = 0$$

Attempt to solve by factoring, but it cannot be factored, so solve using the Quadratic Formula.
The x-intercepts are:

$$x = -1 \pm \sqrt{6}$$

Since these will need to be plotted on the number-line, let's compute the radical and get rounded, numeric values. The square root of 6 is approximately 2.45, so

$x = 1.45$ and -3.45

Plot these on a number-line/1D graph (not shown here). Since the inequality symbol is inclusive, use closed dots or brackets.

Define the intervals using inequalities (since the original inequality symbol is inclusive, make the following inequality symbols inclusive as well):

$x \leq -3.45$, $-3.45 \leq x \leq 1.45$, $x \geq 1.45$

Choose test points for each region...
Let's use: -4, 0, 2

Test each point by substituting it into the original inequality statement.

This is continued on the next page...

Test each point by substituting it into the original inequality statement. I'm going to use the simplified form:

$$x^2 + 2x - 5 \leq 0$$

-4: $(-4)^2 + 2(-4) -5 \leq 0$

$16 - 8 - 5 \leq 0$

$3 \leq 0$ is not a true statement so "x $\leq$ -3.45" is *not* a solution.

0: $(0)^2 + 2(0) - 5 \leq 0$

$-5 \leq 0$ is a true statement, therefore "-3.45 $\leq$ x $\leq$ 1.45" *is* a solution.

2: $2^2 + 2(2) - 5 \leq 0$

$4 + 4 - 5 \leq 0$

$3 \leq 0$ is not a true statement, therefore x $\geq$ 1.45 *is not* a solution.

Looking at the constraints, points must satisfy both "x $\geq$ -3.45" *and* "x $\leq$ 1.45" for the inequality statement to be true. This was proven from test-point 0. Also, since the inequality symbol in the original inequality is "$\leq$ 0", and the parabola is positive, you can assume to use "and."

The solution, in set-builder notation, is:

{x| -3.45 $\leq$ x $\leq$ 1.45} in consolidated form.

In interval notation: [-3.45, 1.45]

CONSTRAINTS with "AND" & "OR"

When you are dealing with certain inequalities, the constraints will play a role in how you set up the problem, and the solution. You should first recall the following statements, as the constraints are often built off of these:

Another way of saying that a quantity is *negative* is by saying it is *less-than zero*.

Another way of saying that a quantity is *positive* is by saying it is *greater-than zero*.

A quantity that is *either negative or zero* is *less-than-or-equal-to zero*.

A quantity that is *either positive or zero* is *greater-than-or-equal-to zero*.

There are a few common constraints that are simply in reference to (the signs due to) the multiplication or division of positive and negative quantities. (I refer to quantities, not just numbers or values, because a quantity includes individual numbers or values, but also includes groups that are not yet simplified, as you will often encounter). This is all based on fundamental principles, just posed in the realm of inequalities. The groups of constraints you are about to see, I call "and-or-and" statements, and I have arranged them accordingly. This is further discussed in: Applying Constraints to the Solution, 4 pages later.

Constraints for Multiplication

Factors multiplied to yield a negative product can be expressed as:
$(+)(-) = -$

This can also be written as:
$(+)(-) < 0$ OR $(-)(+) < 0$

and can even be written as:
(quantity < 0)(quantity > 0) < 0 OR (quantity > 0)(quantity < 0) < 0

InWords: A positive quantity times a negative quantity equals a negative quantity.

The above statement can be summarized as an "and-or-and" statement because, for the inequality statement to be true, either
the left quantity is positive *and* the right quantity is negative
OR
the left quantity is negative *and* the right quantity is positive.

If a product is less-than-or-equal-to zero, it can be setup as:
(quantity ≥ 0)(quantity ≤ 0) ≤ 0 OR (quantity ≤ 0)(quantity ≥ 0) ≤ 0

Factors multiplied to yield a positive product can be expressed as either:
$(+)(+) = +$ OR $(-)(-) = +$

This can also be written as:
$(+)(+) > 0$ OR $(-)(-) > 0$

and can even be written as:
(quantity > 0)(quantity > 0) > 0 OR (quantity < 0)(quantity < 0) > 0

In Words: A positive quantity times a positive quantity equals a positive quantity.
Likewise, a negative quantity times a negative quantity equals a positive quantity.

If a product is greater-than-or-equal-to zero, it can be setup as:
(quantity ≥ 0)(quantity ≥ 0) ≥ 0 OR (quantity ≤ 0)(quantity ≤ 0) ≥ 0

Constraints for Division and Fractions

The signs of the quantities in the numerator and denominator of a fraction affect the sign of the value of the fraction.

A fraction whose value is negative can be expressed as:

$$\frac{(+)}{(-)} = - \quad \text{OR} \quad \frac{(-)}{(+)} = -$$

This can also be written as:

$$\frac{(+)}{(-)} < 0 \quad \text{OR} \quad \frac{(-)}{(+)} < 0$$

and can even be written as:

$$\frac{\text{quantity} > 0}{\text{quantity} < 0} < 0 \quad \text{OR} \quad \frac{\text{quantity} < 0}{\text{quantity} > 0} < 0$$

In Words: A negative quantity over a positive quantity yields a fraction whose value is negative. Likewise, a positive quantity over a negative quantity yields a fraction whose value is negative. This is the same as saying "a negative divided by a positive equals a negative," or that "a positive divided by a negative equals a negative.

A fraction whose value is positive can be expressed as:

$$\frac{(+)}{(+)} = + \quad \text{OR} \quad \frac{(-)}{(-)} = +$$

This can also be written as:

$$\frac{(-)}{(-)} > 0 \quad \text{OR} \quad \frac{(+)}{(+)} > 0$$

and can even be written as:

$$\frac{\text{quantity} < 0}{\text{quantity} < 0} > 0 \quad \text{OR} \quad \frac{\text{quantity} > 0}{\text{quantity} > 0} > 0$$

In Words: A negative quantity over a negative quantity yields a fraction whose value is positive. Likewise, a positive quantity over a positive quantity yields a fraction whose value is positive.

If the fraction can equal zero...
If the value of a fraction is greater-than-or-equal-to zero, it can be set up as:

$$\frac{\text{quantity} \le 0}{\text{quantity} < 0} \ge 0 \quad \text{OR} \quad \frac{\text{quantity} \ge 0}{\text{quantity} > 0} \ge 0$$

If the value of a fraction is less-than-or-equal-to zero, it can be set up as:

$$\frac{\text{quantity} \ge 0}{\text{quantity} < 0} \le 0 \quad \text{OR} \quad \frac{\text{quantity} \le 0}{\text{quantity} > 0} \le 0$$

Notice that in each setup of fractions which could equal zero, the numerators will contain the "or-equal-to zero" option, but the denominators can only be greater-than or less-than zero. They cannot have the option to equal zero, because if and when a denominator equals zero, the fraction becomes undefined.

Binomials in the Denominator
When you are solving a rational inequality, and going through the intermediate simplification steps, you may encounter two binomial factors in the denominator (say, from factoring a quadratic expression). You should treat them as factors multiplied to yield a positive or negative product using the Constraints for Multiplication.

Applying Constraints to the Solution
The statements in this section can all be summarized as "and OR and" statements, for reasons hopefully you have seen in the explanations. But how can these be applied to the solution?

If an "and" statement proves to be true, it can be written in consolidated form. If *both* "and" statements prove to be true, then *both* consolidated statements are the solution, because "or", by definition, means "both" or "doesn't have to overlap."

However, if one of the "and" statements fails (is not true, meaning, no points of overlap), that statement is disregarded. Therefore, the (only) other "and" statement is the solution.

Also, if neither "and" statement proves to be true, there is no solution.

RATIONAL INEQUALITIES

The following are constraints commonly used for Rational Inequalities:
- Involving multiplication
- Involving division and fractions
- A zero in the denominator causes no solution, so any value in the denominator which causes the denominator to equal zero is an extraneous solution[*].

In these types of problems, you must pay very careful attention to the details, particularly the usage of "and" and "or." Rational inequalities can be solved in two ways. One way is by using a number-line and test-points (the 1D Procedure for Solving Quadratic Inequalities), and the other is what I call the "Constraints Procedure." An example will be shown for each procedure.

Constraints Procedure:
1. Rearrange the inequality so that all terms are on the left side of the inequality sign, and zero is on the right.

2. Simplify the left side so there is only one rational expression. You may need to apply the Addition of Rational Expressions[*].

3. Find extraneous solution(s) from the denominator.

4. Is the rational expression less-than or greater-than zero? Follow the appropriate procedure below. The remaining procedures are very similar. They only differ by whether the signs of the numerator and denominator must be the same or opposite.

In short, for the remaining steps, you will solve two complex inequalities with "and" then combine those consolidated statements as an "or" statement.

When the rational expression is **> 0**, either
the numerator *and* denominator must both be positive
or
the numerator *and* denominator must both be negative:
5. Set each the numerator and denominator as > 0, and solve for x in both the numerator and denominator. Write each solution as statements with "and" between them. Write the consolidated solution. Since this is an "and" statement, you must select the

inequality where the overlap occurs. This is the first half of the solution. Save it for Step 7.

6. Set the numerator and denominator as < 0. Solve for x in both the numerator and denominator. Write each solution as statements with "and" between them, then write the consolidated statement which reflects the points of overlap. This is the second half of the solution to use in Step 7.

When the rational expression is ≥ 0, either
the numerator is greater-than-or-equal-to 0 *and* denominator is positive
or
the numerator is less-than-or-equal-to 0 *and* denominator must be negative:

5. Set the numerator ≥ 0, and the denominator > 0, and solve for x in each. Write each solution as statements with "and" between them. Write the consolidated solution. Since this is an "and" statement, you must select the inequality where the overlap occurs. Pay extra attention to the "or-equal-to" point. This is the first half of the solution. Save it for Step 7.

6. Set the numerator ≤ 0 and set the denominator < 0. Solve for x in each. Write each solution as statements with "and" between them, then write the consolidated statement which reflects the points of overlap. This is the second half of the solution to use in Step 7.

When the rational expression is < 0, either
the numerator is positive *and* the denominator is negative
or
the numerator is negative *and* the denominator is positive:

5. Set each the numerator as > 0 and denominator as < 0, and solve for x in both the numerator and denominator. Write each solution as statements with "and" between them. Write the consolidated solution. Since this is an "and" statement, you must select the inequality where the overlap occurs. This is the first half of the solution. Save it for Step 7.

6. You will essentially repeat Step 5, except now, the numerator is $<$ 0 and the denominator is > 0. Solve for x in both the numerator and denominator. Write each solution as statements with "and" between them, then write the consolidated statement which reflects the points of overlap. This is the second half of the solution to use in Step 7.

When the rational expression is ≤ 0, either
the numerator is greater-than-or-equal-to 0 *and* the denominator is negative
or
the numerator is less-than-or-equal-to 0 *and* the denominator is positive:

5. Set each the numerator as ≥ 0 and denominator as < 0, and solve for x in both the numerator and denominator. Write each solution as statements with "and" between them. Write the consolidated solution. Since this is an "and" statement, you must select the inequality where the overlap occurs. This is the first half of the solution. Save it for Step 7.

6. You will essentially repeat Step 5, except now, the numerator is $\leq$ 0 and the denominator is > 0. Solve for x in both the numerator and denominator. Write each solution as statements with "and" between them, then write the consolidated statement which reflects the points of overlap. This is the second half of the solution to use in Step 7.

The remaining steps are the same no matter the original inequality...

7. Take the solutions from Steps 5 & 6 and write "or" between them; this reflects the either-one-*or*-the-other scenario. Note, there is a chance that in Steps 5 & 6, one of the "and" statements has no solutions. That's fine, and you must acknowledge that. In your final solution, just include the "and" statement that works.

8. You may choose to write the solution in consolidated form, and either in set-builder or interval notation. Remember to exclude any extraneous solution from the solution from Step 3, if necessary.

9. You also may be required to graph the solution on a number-line. Remember to use the proper inclusive/noninclusive symbols. Since the final statement is an "or" statement, arrows on the number-line should be pointed away from each other, with a space between them.

Examples begin on the next page...

Example 1:

$$\frac{x+1}{x^2 - x - 6} \geq 0$$

The inequality is already arranged so all terms are on the left and zero is on the right.
Also, there is only one rational expression on the left, so no extra work needs to be done in that respect.

Find the extraneous roots of the denominator. It can be factored into:
$(x - 3)(x + 2)$

Set each parenthetical group equal to zero and solve for x to find the extraneous solutions of
$x = 3$ and $x = -2$

Since the inequality is ≥ 0, either the top *and* bottom must both be positive, *or*
the numerator is less-than-or-equal-to 0 *and* denominator must be negative.

Set the numerator ≥ 0, and the denominator > 0, and solve for x in each. Start with the numerator:
$x + 1 \geq 0$

Therefore, $x \geq -1$

In the denominator (remember, the denominator can never be "or-equal-to zero"):
$(x - 3)(x + 2) > 0$

Each parenthetical group must be > 0, so

$x - 3 > 0$ and $x + 2 > 0$

After solving,

$x > 3$ and $x > -2$

Continued on the next page…

Define the area of overlap, which is $x > 3$. Combine this with the numerator:

$x \geq -1$ and $x > 3$

Choose the area of overlap, which is $x > 3$. This is half of the solution.

Now set the numerator ≤ 0 and set the denominator < 0
The numerator becomes:

$x \leq -1$

The denominator becomes:

$x < 3$ and $x < -2$

The area of overlap is $x < -2$

Now choose the area of overlap of the numerator and denominator. It is:

$x < -2$

The solution is:

$x < -2$ or $x > 3$

In consolidated form: $-2 > x > 3$

In interval notation: $(-\infty, -2) \cup (3, \infty)$

Example 2:

Solve the following inequality using a number-line and test-points:
$$\frac{2x - 2}{x + 3} < 1$$

Move the 1 to the left by subtracting it from both sides, so 0 is on the right.
$$\frac{2x - 2}{x + 3} - 1 < 0$$

In order to have only one fraction on the left, convert the 1 to:
$$\frac{x + 3}{x + 3}$$

and subtract it from the other fraction:
$$\frac{2x - 2}{x + 3} - \frac{x + 3}{x + 3} < 0$$

Distribute the negative sign through the numerator terms:
$$\frac{2x - 2 - x - 3}{x + 3} < 0$$

Combine like-terms:
$$\frac{x - 5}{x + 3} < 0$$

Find the x-intercepts for the numerator and denominator. They are:
x = 5 and x = -3
Plot these on the number-line. Define each region. They are:
(-∞, -3), (-3, 5), (5, ∞)

Choose test-points for each region. Let's use -4, 0, and 6.

Substitute these into the simplified inequality. Start with -4:

$$\frac{-4 - 5}{-4 + 3} < 0$$

$$\frac{-9}{-1} < 0$$

9 < 0 is not a true statement, so region (-∞, -3) is rejected.

0:

$$\frac{0 - 5}{0 + 3} < 0$$

$$\frac{-5}{3} < 0$$

This point proves true, so region (-3, 5) is accepted.

6:

$$\frac{6 - 5}{6 + 3} < 0$$

$$\frac{1}{9} < 0$$

This point is not true, so region (5, ∞) is rejected.

The solution in interval notation is: (-3, 5).

Note that the extraneous solution of "-3" does not conflict with this since the graph approaches but never touches -3.

The solution in set-builder notation is {x| -3 < x < 5}.

VARIATION & PROPORTIONALITY

Variation refers to the relationship of a variable with respect to some other quantity. That *quantity* may be a single variable, a variable raised to an exponent, the root of a variable, or a group of variables, in the numerator or denominator. A common question is:
"As the value of one variable goes up, what happens to the value of another related variable or quantity... Specifically, does it proportionally go up or down?"

Variation and proportionality are based on (maintaining) the *proportionality constant (a.k.a. the variation constant).* Variation refers to how one variable is affected by the change of the value of another related quantity. Proportionality refers to: if and how one variable maintains its proportionality with respect to another variable or quantity, across a set of data points. Although "variation" and "proportionality" are defined with slightly different slants, they are essentially synonymous and can be (and are) used interchangeably.

Let's examine this concept from the beginning. Suppose you have a series of data points (let's call them ordered pairs of x & y), then you go on to test if there is a consistent trend between them. For each point, you make a ratio (fraction) of y vs. x as x over y. For each point, the ratio of x over y equals the same number (call this "k" which will be explained shortly). Because the ratio of each point equals the ratio of every other point, they are proportional, and thus x and y are proportional. The x-value and the y-value for each point probably do not equal each other though, but since they are mathematically related in a consistent, predictable pattern, they are set across from each other as:

$$y \propto x$$

where $\propto$ is the "proportionality symbol." This is not in equation form, but it can be converted to equation form by replacing the proportionality symbol with "= k":

$$y = kx$$

Since the variation constant is a fixed value, it can be used with a known x-value to solve for the related, unknown y-value, or vice-versa.

Direct variation (a.k.a. directly proportional) means: *As the value of one variable increases, the value of a related quantity increases proportionally.* Likewise, if the value of a variable decreases, the value of a related quantity decreases proportionally. In this case, the value of k is > 1.When a variable and a quantity *directly vary*, the variable and the related quantity are both in the numerator, on opposite sides of the equation in the form of:

y = k(quantity)

If the equation is rearranged (solved for k) so that the variable and the related quantity are on the same side of the equation, the variable and the related quantity will be in opposite parts (top and bottom) of the fraction, in the form of:

$$k = \frac{y}{\text{quantity}}$$

Inverse variation (a.k.a. inversely proportional) means: *As the value of one variable increases, the value of a related quantity decreases proportionally.* Likewise, if the value of a variable decreases, the value of a related quantity increases proportionally. In this case, k has a fractional value (0 < k < 1). When a variable and a related quantity inversely vary, the variable and the related quantity are in opposite parts of a fraction when they are on opposite sides of the equation, in the form of:

$$y = \frac{k}{\text{quantity}}$$

If the equation is rearranged (solved for k) so the variable and related quantity are on the same side of the equation, they will be factors in the same part of the fraction, in the form of:

k = y(quantity)

Note: Some people incorrectly call "inverse variation" "*indirect* variation," but this is incorrect. Although it is the opposite of *direct* variation, the true opposite is *inverse* because of the positioning of values in the equation.

There are a number of common misconceptions people have on this subject. In order to prevent them, it is best to take a closer look at this from a proportionality perspective.

More on Proportionality

When a set of data which has proportionality is graphed, it produces a straight line. You can relate this to the slope-intercept form of a linear equation $y = mx + b$, where m is the proportionality constant (and the y-intercept can be anything; it doesn't affect the proportionality). Only a straight line can be produced by data which is proportional (or varies directly). There is an important reason for stressing this point.

You will likely be asked to examine equations in which some variable (say, y) varies directly with, say "x^2" from an equation such as $y = kx^2$. The graph from this equation will be a parabola, not a straight line, so it's important to understand that the proportional relationship is between y and the quantity x^2, not x. When values of y are plotted vs. values of (x^2), this will produce a straight line as a result of a proportional relationship. This is why I often refer to a "quantity," not just a variable, in this chapter, because by saying "a variable varies directly with another variable," can be misleading for certain equations. Another reason I refer to a quantity is because a variable can be related to a combination of variables, which is called *joint variation*.

The graph of inverse variation, however, is not a straight line. It produces a curved line with a negative trend.

CIRCLES

When the center of a circle is the origin, the equation is:
$$x^2 + y^2 = r^2$$

This is the parent equation, based on the parent graph.
Equations for circles are all built around the above equation and the shift of the center point from the origin. The shift of the center from the origin is reflected by "h" and "k".

h is a specific x-value showing the horizontal distance (shift) that the center point is from the origin.
k is a specific y-value showing the vertical distance (shift) that the center point is from the origin.

Since h and k are both zero in the parent graph, you can also write it as:
$$(x - 0)^2 + (y - 0)^2 = r^2$$

The Standard Form Equation for a Circle is:
$$(x - h)^2 + (y - k)^2 = r^2$$

What are h and k?

(h, k) represent the center point of a circle on a graph.
- Sometimes you are given an equation and asked to extract the center point of the circle.
- Or, you may be given the graph and asked to write the equation.
- (You might also be given a non-standard form equation and asked to make it into standard form. This is discussed below.)

Note: In the Standard Form Equation for a Circle, the "h" and "k" are *subtracted* from x and y, respectively, in the parentheses. This is independent of the signs of the actual values of h and k, which might be negative. It is important that you don't let this confuse you. This is explained in a way to help you assimilate it to other topics you may already know in:
The Similarity of The Pythagorean Theorem, The Distance Formula, & The Circle Equation.

I want you to recall and connect this to Solving a Quadratic Equation by Factoring[*]. Isn't it true that when doing this method, you factor into parenthetical groups, then set the contents of each group equal to zero, and solve for x?

Suppose you had: $y = x^2 + x - 12$

It would factor into: $(x - 3)(x + 4)$

Set $x - 3 = 0$, so $x = +3$, and
Set $x + 4 = 0$, so $x = -4$

The point is that the sign of the answer is opposite of the sign as it was in the parentheses. You probably thought nothing of this when solving a quadratic equation by factoring.

r represents the radius of the circle – the distance from the center of the circle to any point on the curved line of the circle.
"r" is specific to circles, but in a more general sense, it is analogous to
- "c" from the Pythagorean Theorem, and
- "d" from the Distance Formula.

r^2 is simply the value of the radius-squared. This is the value on the right side of the Standard Equation for a Circle, however this value is not present on the graph. If you have this value (from the equation), you can find the radius, r, by taking the square root of the r^2 value.

You will likely be given a few starting bits, and be told to make the Standard Form equation for a Circle, which will include:
- a binomial squared of x, plus
- a binomial squared of y, equal to
- a constant, which is the radius-squared

It will ultimately look something like:

$(x +/- \#)^2 + (y +/- \#)^2 = $ some #

This procedure is based on using the Complete the Square Method twice, once for the x's and once for the y's, independent of each other.

Procedure for Making the Standard Form Equation of a Circle

1. Rewrite the equation, as shown below, moving the constant (the number with no variable) to the right side of the equal-sign, and leaving a distinct space after the x terms, a space after the y terms, and two spaces on the right side, after the constant.

$$x^2 + \#x + \underline{\quad} + y^2 + \#y + \underline{\quad} = -c + \underline{\quad} + \underline{\quad}$$

2. You will perform the Complete the Square Method for the x-terms and the y-terms, which will create 2 trinomials on the left:

    2a. Complete the square for the x-terms by finding the *new* c_x, and write it in the designated space on the left, and on the right side.

    2b. Complete the square for the y-terms by finding the *new* c_y, and write it in the designated space on the left, and on the right side.

It should now resemble:

$$x^2 + \#x + \textit{new } c_x + y^2 + \#y + \textit{new } c_y = -c + \textit{new } c_x + \textit{new } c_y$$

3. Convert the perfect square trinomials you just "completed" into binomials-squared, and on the right, combine like terms into one number (which now represents the radius-squared). It should look like:

$$(x +/- \#)^2 + (y +/- \#)^2 = \# \leftarrow r^2$$

This equation is now in Standard Form for a Circle.

Graphing a Circle

0. If the equation is not in Standard Form, you will need to go through the procedure to convert it, in order to perform the next steps.

1. Identify the Center Point: (h, k).

2. Determine the radius, r. You may need to take the square root of the "r^2" number from the Standard Form equation.

3. Plot the center point from Step 1.

4. Using the center point as a reference, and the value of "r", plot 4 more points:
- a point "r" units directly *above* the center point,
- a point "r" units directly *below* the center point,
- a point "r" units directly to the *right* of the center point, and
- a point "r" units directly to the *left* of the center point.

5. Sketch the circle by connecting the four outer dots with a smooth, curved line (so as not to make a diamond).

6. It's a good idea to label your center point.

THE SIMILARITY OF THE PYTHAGOREAN THEOREM, THE DISTANCE FORMULA, & THE CIRCLE EQUATION

This section is designed to show you (and prove to you) how the Distance Formula and the Standard Form Equation for a Circle are other forms of the Pythagorean Theorem, an equation you are likely familiar with. As you know, the philosophy of *Algebra in Words* is to connect topics that a textbook might not.

The Pythagorean Theorem is:
$$a^2 + b^2 = c^2$$

The Distance Formula is:
$$\sqrt{(x_2 - x_1)^2 + (y_2 - y_2)^2} = d$$

The Standard Form Equation for a Circle is
$$(x - h)^2 + (y - k)^2 = r^2$$

Let's look at how they relate by looking at each equation in more detail…

The Pythagorean Theorem is:

$a^2 + b^2 = c^2$ for all right-triangles. In a geometric and trigonometric sense:

- Each side of a triangle is associated and named (with a letter) by the angle it is across from.
- Angles are given capital letters and
- The *lengths* of the sides are named with lower-case letters.
- Each side of a triangle is also given a whole-word name in reference to the smallest angle and the right (90 degree) angle…

Angle "A" is the smallest angle[†]. So,
"a" is the length of the side opposite that, called the "opposite" side.

Angle "B" is the middle-sized angle[†]. So,
"b" is the length of the side across from it. But it is named the *adjacent* side. Adjacent means "next to," and "b" is always *adjacent* to angle A (the smallest angle). The other side adjacent to angle A is…

The *hypotenuse*. The hypotenuse is the side across from the right-angle, which is always angle C.
"c" represents the length of the hypotenuse.

[†]Note: What about for a right isosceles triangle, one in which two angles are the same and two sides are the same length? In this case, you can call either of the non-hypotenuse sides "a" and "b".

To solve for c from the Pythagorean Theorem,
$a^2 + b^2 = c^2$

you take the square root of both sides, giving you c:
$$\sqrt{a^2 + b^2} = \sqrt{c^2} = c$$

$$c = \sqrt{a^2 + b^2}$$

Note: Although the Square Root Property tells you that taking the square root of a number will give you a positive and a negative answer, you only accept the positive value for lengths, because lengths can only be positive.

The Distance Formula is:

$$\sqrt{(x_2 - x_1)^2 + (y_2 - y_2)^2} = d$$

Suppose we square both sides, to look at it from that perspective. It would look like:

$$(x_2 - x_1)^2 + (y_2 - y_1)^2 = d^2$$

Doesn't this look familiar to the Pythagorean Theorem? Remember that each group, and d^2, represent real-number values, when numbers are substituted in for the variables.

To solve for "d" you would take the square root of both sides:

$$\sqrt{(x_2 - x_1)^2 + (y_2 - y_2)^2} = \sqrt{d^2} = d$$

which takes you back to the Distance Formula.

Now look at the Standard Form Equation of a Circle:

$$(x - h)^2 + (y - k)^2 = r^2$$

Doesn't this look like the Distance Formula with both sides squared? Now look at the Standard Form Equation of a Circle when the center point is the origin, when h and k are both 0:

$$x^2 + y^2 = r^2$$

Doesn't this look like the Pythagorean Theorem, just with different letters?
Here is how they are related:

Think of a circle sketched on a graph, with the generic center (h, k).
1. From the center point, draw a straight, diagonal line to any point on the circle; this would be the radius, with distance and symbol "r" but let's prove it.
2. From that point on the circle, sketch a line parallel with the y-axis, until it becomes even with the radius.
3. Fill in a dot as a point, generically called (x, y), being that it is x units, horizontally, from the origin, and y units, vertically, from the origin.
4. From point (x, y), draw a line, so it is parallel with the x-axis, back to the center of the circle. You just made a right triangle, which the Pythagorean Theorem can be applied to.
5. The length of the horizontal side could be found by "x – h".
6. The length of the vertical side could be found by "y – k".

7. Looking at the triangle on the graph, r is the equivalent of c. The horizontal and vertical sides are the equivalent of "a" and "b". Substituting these into the Pythagorean Theorem, you can see where The Standard Form Equation of a Circle,

$$(x - h)^2 + (y - k)^2 = r^2$$

comes from. Taking the square root of both sides, as you would to find c in the Pythagorean Theorem, allows you to find the distance of the radius, r:

$$\sqrt{(x - h)^2 + (y - k)^2} = \sqrt{r^2} = r$$

This is discussed more in: Circles.

Question for thought: If $a^2 + b^2 = c^2$, does $a^3 + b^3 = c^3$?
Does the equation work for any power other than "2" ?
This is known as Fermat's Last Theorem. Fermat, in short, postulated: No, it does not, however, he died before fully publishing his proofs. It has since been explored by other mathematicians.

SCIENTIFIC NOTATION: WHICH WAY DO YOU MOVE THE DECIMAL?

Scientific Notation is a form of expressing a number with many placeholders in it (such as zeros and decimal places). Scientific Notation is a way of converting numbers (which are often either very large or very small in value) from normal form (a.k.a. *expanded form*) into a more manageable form, while still maintaining the value and accentuating the significant digits. Scientific notation contains two general parts: the coefficient factor and the base-ten raised to an exponent factor:
coefficient x $10^{exponent}$

More specifically, the proper form of scientific notation, from left to right, consists of:
One non-zero digit, then the decimal, the remaining significant digits, times ten to the exponent, written like:
$\#.\#\#\# \times 10^{\#}$

The problems you will be given will usually tell you how many significant digits to have in the coefficient, and you will likely have to round the last (right-most) digit if you had more digits to start with in the original (expanded) form. *Significant digits* are the number of digits meant to be in the coefficient. The digits after the decimal may be either zero or non-zero digits. Even though we generally stop using the "x" for "times" once we graduate from arithmetic, it is still often used in scientific notation, so do not confuse it with a variable. That being said, sometimes parentheses are used for "times" as:
$\#.\#\#\# (10^{\#})$

A few examples of common numbers in scientific notation are:
- The speed of light: c = approximately 3.00×10^8 meters/second
- Avogadro's number: $N_A = 6.022 \times 10^{23}$ particles/mole
- The sun is 9.296×10^7 meters from Earth
- A red blood cell is about 7×10^{-6} meters in diameter

Scientific notation is often used for expressing and converting measurement units in the metric system because scientific notation and the metric system are all based on factors of 10 (a.k.a orders of magnitude).

You can convert an expanded-form number into scientific notation, and a number in scientific notation into expanded form.

A number in scientific notation with a negative exponent signifies that the original number in expanded form is a smaller number.

A number in scientific notation with a positive exponent signifies that the original number in expanded form is a larger number.

Expanded Form to Scientific Notation

Often, for non-decimal numbers, the decimal is not written at the end of the digit(s), but it is still considered to be there. For instance the number seven is usually written as "7" but is essentially the same as "7.0" and can even be written as "7." with the decimal at the end with no zeros written after it. Anyway, this (the placement of the decimal) is your starting point.

Count from where the decimal *is* to where it's supposed to *go*, counting (between) one digit at a time, in the proper direction, until you land exactly to the right of the left-most non-zero digit. I recommend doing this by drawing little arrows with your pencil, while you count the spaces.

Write the new form of the number. If you are counting significant digits, they are counted from left to right, starting with the one digit to the left of the decimal. If you moved the decimal from right to left, your exponent will be positive, as many spaces that you moved the decimal. If you moved the decimal from left to right, your exponent will be negative, as many spaces as you moved the decimal.

Example:
Suppose 120,452,337 people watched the Super Bowl. Convert this number into scientific notation with three significant digits.

The decimal (although not written) is to the right of the 7. Move it to the right of the "1", which requires a movement of 8 places left. Since we moved the decimal left, the exponent will be positive. Since it must be in three significant digits, keep the 2 and 0 (you already have the 1). Also, look at the digit to the right of 0... since 4 is "a number below 5," the zero does not need to be rounded up to 1; it stays 0. This becomes:
1.20×10^8

Example:
Convert 0.0000053499 into scientific notation with two significant digits.

The decimal goes to the right of the 5, and is moved 6 places to the right. Since it must be in two significant digits, look to the 4 to the right of the 3. Since that 4 is less than 5 (despite the 9s to the right of the 4), the 3 is left as 3 and not rounded up. Since the decimal was moved to the right, the exponent will be negative. This becomes:
5.3×10^{-6}

Question: Is 56.87×10^4 in scientific notation?
Answer: Technically, no, because there is not only one digit to the left of the decimal. That being said, numbers are permitted to be written this way, and maintain their value, they just aren't in scientific notation.

Scientific Notation to Expanded Form

The sign of the exponent (of base-ten) is the instructions on how to convert a number from scientific notation to expanded form.

If the exponent is positive, move the decimal to the right that many places.

If the exponent is negative, move the decimal to the left that many places.

Example 1:
Convert 1.892×10^4 to expanded form.
Since the exponent is positive 4, move the decimal four places to the right. Since the 8, 9 and 2 constitute three places, fill in the fourth place with a zero. It is converted to: 18,920

Example 2:
Convert 1.5×10^{-9} to expanded form.
Since the exponent is negative 9, move the decimal 9 places to the left. Keep the digit 5 attached. Also, we tend to write a zero to the left of a decimal if there is no other digit to the left of the decimal. It is converted to: 0.0000000015

Note on interpreting the exponent:
A negative exponent tells you how many *decimal places* there are to the left of the first non-zero digit. It *does not* indicate how many *zeroes* are to the left between the decimal and the first non-zero digit. This is a false assumption and FMM some students make. Actually, there will always be *one fewer zeros* than the number of decimal places to the left of the first non-zero digit. Observe in the previous example that the exponent was negative nine, but there are 8 zeros to the left of the 1, in expanded form.

Example 3:

Simplify the following fraction, not by expanding the numbers in scientific notation and computing, but by treating the numbers as coefficients and like-bases-of-ten, and applying the principle of multiplying and dividing like-bases, and using the commutative property of multiplication:

$$\frac{(2.0\ x10^5)(3.0\ x10^{-3})}{(3.0\ x10^{-2})}$$

Rearrange this so the coefficients are next to each other and the base-tens are next to each other:

$$\frac{(2.0)(3.0)(10^5)(10^{-3})}{(3.0)(10^{-2})}$$

On the top, multiply the coefficients 2.0 and 3.0, then combine 10^5 and 10^{-2} by adding their exponents. The bottom stays the same in this step:

$$\frac{(6.0)(10^2)}{(3.0)(10^{-2})}$$

Now divide coefficients and base-tens in the top and bottom to get:

$2.0\ x10^2$

This example is to demonstrate that doing computations with scientific notation are possible, and might even be simple, if you don't have your calculator.

LOGARITHMS & NATURAL LOGS

A logarithm is the inverse function of an exponential function. Thus far in algebra, you have been exposed to many examples of solving for an unknown variable… when the variable is the base. But what about solving for variables in the exponent?

The *property of exponential equations* says that you can solve for a variable in the exponent as long as you can set one known-base-raised-to-a-*known*-exponent equal to the same known-base raised to the *unknown* exponent (variable), as:

$b^x = b^y$,   where b (for base) is any positive number except 1.

But what if the bases aren't known, aren't the same, or can't easily be factored into exponential form? That's where logarithms come in. You can't solve for an unknown variable in the exponent using traditional algebraic methods, but logarithms provide an easy way to solve for a variable in the exponent because of the way logs can be rearranged and manipulated.

The equation based on the definition of a logarithm is:

$y = \log_b x$ (*In Words*: y equals log base b, of x)

which is a rearrangement of (or can be rearranged into):

$x = b^y$

where x is a real number and y is the (usually unknown) variable in the exponent to be solved for. It may be easier to view it as:

unknown variable exponent $= \log_{base} \#$

which could be rearranged into:

$base^{unknown\ variable} = \#$

One possibly confusing aspect of log equations is when the unknown variable symbol is "y", when it might seem more logical to think of it as $x = \log_b y$, because you are used to solving for x as the unknown, independent variable. At the end of the day, though, these variables are arbitrary, and will often be filled-in with either a number (for known x), or nothing (for known y) until you solve for it, which will again be a number. I believe it is written as:

"$y = \log_b x$"

so y can be substituted for f(x), and the other side is a function of x, as:

$f(x) = \log_b x$.

That being said, sometimes it is written as "$x = \log_b y$".

Converting the log form into (traditional) exponential equation form can help you work towards solving for the unknown variable in the exponent if you can convert the "#" (by factoring it) into exponential form with the property of exponential equations. See Example 1, 8 pages ahead:

But what about when the # can't be easily factored to have the same base? Again, that is where logarithms become useful, and the following properties help you get around that obstacle.

Important Note: From my research, I couldn't find any books which name (what I call) "#" above (although, sometimes they are called "real numbers," but this is still too ambiguous). Since all other parts of a log have a name, and since this book is about being *in words*, continuing on, I will refer to # as "the quantity," for lack of a better name. This *quantity* is the # represented by "x" in the above definition, whether it is in the *exponential equation* or in the *log form equation*. "Quantity," in this context, may be a number, variable, or even a polynomial.
Also, when I say "compute," I mean: type it into your calculator and solve to get a number.

LOG PROPERTIES

Note: The following properties (of a product, a quotient, and a power) all can only be performed with *like-bases*.

Power Property:

$$\log_b M^{\text{power}} = (\text{power}) \log_b M$$

$$\log_b M^{\text{the coefficient-to-be}} = (\text{coefficient}) \log_b M$$

In Words: The log whose quantity is raised to a power can be rearranged (into *expanded form*) and rewritten so the power of that quantity is moved as a coefficient to the log. This is multiplied times the log and any coefficients already in front of the log. However, if the power (moved as a coefficient) is a variable, it will not actually be multiplied, it will act as a factor until further equation rearrangement facilitates the solving of that variable, as seen in Step 3 of the Procedure for using Logs to solve for an unknown variable in the exponent.

Note: This is different than, and thus does not apply to, a log raised to a power: $(\log_b quantity)^{power}$

Log of a Product:

$$\log_b MN = \log_b M + \log_b N$$

In reverse:

$$\log_b(\text{quanity } 1) + \log_b(\text{quantity } 2) = \log_b[(\text{quantity } 1)(\text{quantity } 2)]$$

In Words: The log of a product can be split apart into (*expanded form* as) the sum of the logs~of the same base~ of each factor. Likewise, the sum of logs~of the same base~ of given quantities can be consolidated into the log~of the same base~ of the product of those quantities. However, the multiplication of logs cannot be rearranged into any other form, even if their bases are the same:

$$(\log_b M)(\log_b N) \neq \log_b MN \text{ or } [\log_b M + \log_b N]$$

Log of a Quotient:

$$\log_b \frac{M}{N} = \log_b M - \log_b N$$

$$\log_b \left(\frac{\text{numerator}}{\text{denominator}}\right) = \log_b \text{numerator} - \log_b \text{denominator}$$

In reverse:

$$\log_b \text{quantity} - \log_b \text{quantity} = \log_b \left(\frac{\text{quantity of minuend}}{\text{quantity of subtrahend}}\right)$$

In Words: The log of a fraction can be split apart into (*expanded form* as) the log of the numerator minus the log of the denominator. Likewise, the log~of a common base~ of a quantity minus the log~of the same base~ of a quantity can be consolidated into the log~of the common base~ of the (minuend quantity over the subtrahend quantity). However, the division of logs cannot be converted into the forms shown here:

$$\frac{\log_b M}{\log_b N} \neq \log_b \frac{M}{N} \text{ or } [\log_b M - \log_b N]$$

but there is an instance where $\frac{\log_b M}{\log_b N}$ fits into another conversion…See: Change of Base Formula, 2 pages later.

Common Log:

$Log_{10}(quantity) = log\ (quantity)$

In Words: A log whose base is 10 is called a *common log*. When no number for the base is written, it is assumed to be "10". So,

$y = log\ \#$ can be rearranged to: $10^y = \#$

This is the essence of the anti-log. Also, common logs are typically the function on calculators, used to compute constants (numbers).

Natural Log:

When the base of a log is "the number e", this becomes its own subcategory of logs, known as the Natural Log, written as "ln". (If you are wondering why the abbreviated symbol is ln and not nl, it is because the name for natural log is rooted in Latin, and in Latin, the adjective is placed after the noun it describes).

$Log_e(quantity) = ln\ (quantity)$

$[y = log_e x] = [y = ln\ x]$

$y = ln\ x$

Change of Base Formula using Common Logs:

A log whose base is not 10 or e can be solved by the following relationship and rearrangement:

$$\log_b x = \frac{\log x}{\log b}$$

or

$$\log_b x = \frac{\ln x}{\ln b}$$

In Words: When you are taking a log base b of some quantity, this can be rearranged into a ratio where the top log is of quantity x, the bottom log quantity is of b (the base from the left log), and the bases of the log in the top and bottom are the same as each other, but different than the base on the left. In this setup, b and x are positive numbers.

Note: This rearrangement is usually done when the original base is neither 10 nor e because when it (the base) isn't, it isn't as easy to compute (either manually or on a calculator). The original base could be 10 or e, though, but if it is, there's usually no reason to do the rearrangement because it could easily be computed as is. Also, by definition, the logs in the ratio must be of the same base, but the most useful approach is to make the base either 10 (thus, common logs), or e (in which case you're technically taking the ln of x in the top and the ln of b in the bottom).

Logs of 1, 0, & Negatives:

$\text{Log}_{\text{any base}} 1 = 0$
$\text{Log } 1 = 0$
$\text{Ln } 1 = 0$

In Words: The log, any base, of one is zero. Therefore, common log of 1, and natural log of 1 are both zero. This is because, graphically, the x-intercept of this function is 1. The x-intercept may be different, though, if the log equation is altered, causing a horizontal or vertical shift.

$\text{Log}_{\text{any base}} 0$ is undefined. Therefore,
$\text{Log } 0$ is undefined, and
$\text{Ln } 0$ is undefined.

In Words: The log, any base, of zero is undefined. This is best understood because the graph of this function does not cross though any point where $x = 0$ (a.k.a., the y-axis), which is a log-graph's asymptote. A graph may cross the y-axis, though, if the log equation is altered, causing a horizontal shift.

$\text{Log}_{\text{any base}}$ (any negative number) is undefined
Ln (any negative number, or zero) is undefined

In Words: The log, any base, cannot be taken of a negative number. This is best understood because the graph does not cross any points in which x is a negative number (the graph never goes to the left of the y-axis. A graph may have negative x-points (to the right of the y-axis), though, if the log equation is altered, causing a horizontal shift.

Inverse Log (Taking log base b… of b):

$\text{Log}_b b^x = x$

$\text{Log}_b b = 1$

$\text{Log } 10^x = x$

$\text{Ln } e^x = x$

In Words: When the log of a base is taken of a quantity with the same base, the result is the exponent of that quantity. Suppose you took the four Inverse Log statements above, and converted them into exponential form… This is the essence of the Anti-log, which is specific to base-ten, as well as the Inverse Natural Log being specific to base $\underline{e}$.

Antilog:

Go from "$y = \log x$" to $[10^y = 10^{\log x}]$
which becomes 10^y, or some $\# = x$

In Words: To eliminate an isolated log in an equation, take base-ten raised to both sides. The left side stays ten-to-the-power, but on the other side, the "$10^{\log}$" part is crossed out leaving only x. This is typically performed when y is a (known) constant because taking base-ten raised to that constant will result in a number. On a calculator, this function usually appears as "$[10^x]$" or "$[10^\wedge]$" or even "[Antilog]", and is usually the "2nd" or "alt" function of the "log" button. Be sure not to confuse this with the button used for scientific notation, which might look something like "$[x10^\wedge]$" (for *times ten to the*), and is a little different.

Inverse Natural Log:

Go from "$y = \ln x$" to $[e^y = e^{\ln x}]$
which becomes e^y, or some $\# = x$

In Words: To eliminate the "ln", take base-e raised to both sides. On a calculator, this function usually appears as "$[e^x]$", and is usually the "2nd" or "alt" function of the "ln" button. Be sure not to confuse this with the button for scientific notation, which is sometimes either "[e]" or "[EE]"

Taking the Log of Both Sides – The Procedure for finding an unknown variable in the exponent, given in exponential form, using Common Logs:

1. Arrange the equation to isolate the term-with-the-variable-in-the-exponent on one side (the left), if you can. If you can't yet, that's ok. Continue to Step 2.
2. Take the common log of both sides. You now have a Logarithmic Equation.
3. Move the exponent down as the coefficient to the log (using the Power Expansion Property). If the exponent is a polynomial, you must:
   - Distribute the log to each term in the polynomial group. Each term will be a coefficient to the log; one coefficient may be a number and one may be a variable. Isolate the log term with the variable on the left and move the log term with the constant to the right.

At this point, if you want to get an answer in *log (exact) form*:
   - Rearrange without computing.

If you want to get an *approximated number*, you can either:
   - Compute, then divide, or
   - Divide, compute, then divide.

If you: Compute, then Divide:
4. On the left, compute the log of the number.
5. On the right, compute the log(s) of the number(s). Combine if necessary.
6. Divide both sides by the number from the left, which will isolate the variable on the left and will equal the number on the right.

Alternatively, you can Divide, Compute, then Divide (starting after Step 3):
4. Divide both sides by "log #" (without computing yet).
5. On the right, now that a fraction (division) is set up, compute the top logs, combine if necessary, compute the bottom log, then divide their numbers. This will give you the solution of the isolated variable on the left set equal to the number on the right.

See this procedure demonstrated in Example 3, 3 pages ahead.
Note: This would also work by taking the Natural Log of both sides instead of the Common Log.

Procedure for Solving Logarithmic Equations using Log Properties in which x is in the log quantity, not in the exponent:

0. Simplify inside the quantity groups, if possible.

1. Position the equation so there is a constant on one side (the right), and all logs of quantities with variables are on the other side (the left). If there is no stand-alone constant to start, but there is a common log of a constant, isolate it on the right; you can either compute it now, or do it later (or not at all, if you want your answer in *exact* form).

2. On the left, consolidate the logs into one log according to the log properties. (Double-check that the logs to be combined are of the same base). This must be done to facilitate the next step. Move any coefficients to the exponent of the quantity, first, before consolidating the sum or difference of logs.

3. What is the base of the log? Take that base and raise it to both sides of the equation. If it is a common log, make the base 10. If it is a natural log, make the base "e".

4. On the left, cancel out the $\log_b$, leaving only the quantity.

5. On the right, compute the base to that constant, yielding a new constant. This should now be set up as a traditional algebraic equation.

6. Solve for x using algebra[*]. Remember to find extraneous solutions if you have a rational equation.

7. Check your solution. Remember that the log of a negative quantity is undefined, so any value of x which makes that quantity negative must be disregarded. Also, be sure to disregard any extraneous solutions you found from Step 6.

See this procedure demonstrated in Example 4, 3 pages ahead.

Example 1:

Solve for y in:

$y = \log_2 8$

First, rearrange it into an exponential equation:

$2^y = 8$

Factor the 8 into exponential form such that the base is the same as base 2, which is raised to the unknown. This is accomplished by changing eight to two-cubed:

$2^y = 2^3$

Now, you can set the exponents equal, and it's obvious that $y = 3$.

Example 2:

Use the Change Base Formula with Common Logs to solve for y in:

$y = \log_2 8$

Set $\log_2 8 = \dfrac{\log 8}{\log 2}$

Notice the top and bottom logs are common logs, meaning they have the common base 10. Compute the top and bottom, each, using a calculator.

$$\dfrac{\log 8}{\log 2} = \dfrac{0.903}{0.301}$$

Divide the numbers:

$\dfrac{0.903}{0.301} = 3,$ therefore: y, and $\log_2 8 = 3$

Example 3:

Take the Log of Both Sides to solve the following equation in which the unknown variable is in a binomial in the exponent:
$7^{x+3} = 12$

Notice that the bases 7 and 12 are different and can't easily be factored into like-bases. That is why we:
Take the common log of both sides:
$\log 7^{x+3} = \log 12$

On the left, move the exponent as a coefficient to the log:
$(x + 3)\log 7 = \log 12$

Distribute $\log 7$ through $(x + 3)$. This breaks out the variable and separates it from the constant (3):
$x\log 7 + 3\log 7 = \log 12$

Isolate the log-term with x by subtracting $3\log 7$ from both sides, moving the constant to the right:

$x\log 7 + 3\log 7 - \textbf{3log } \textbf{7} = \log 12 - \textbf{3log } \textbf{7}$

$x\log 7 = \log 12 - 3\log 7$

You could compute, rearrange, then divide. But let's rearrange, then compute, then divide. Divide both sides by log 7, which isolates x:

$$x = \frac{\log 12 - 3\log 7}{\log 7}$$

Leaving the answer this way is in *log (exact) form.* Let's proceed to get a number. The following computations are rounded to 3 decimal places:

$$x = \frac{1.079 - 3(0.845)}{0.845} = \frac{1.079 - 2.535}{0.845}$$

$$x = \frac{-1.456}{0.845}$$

$x = -1.723$

This answer is in *number (approximate) form.*

Example 4:

Solve for x in the following logarithmic equation, in which x appears in more than one log quantity (but not in the exponent):

$\log_2 x = \log_2 (x + 5) - 1$

The quantities are already simplified.
Rearrange the equation so the constant "-1" is on the right, and the other terms are on the left, by subtracting "$\log_2 (x + 5)$" from both sides:

$\log_2 x \textbf{- log}_2 \textbf{(x + 5)} = \log_2 (x + 5) - 1 \textbf{- log}_2 \textbf{(x + 5)}$

$\log_2 x - \log_2 (x + 5) = -1$
Note: "$\log_2 (x + 5)$" cannot be expanded by any principle; it's ok that the "5" is inside the quantity right now. This log will be consolidated with "$\log_2 x$" by the Quotient Property (going from expanded to consolidated form):

$$\log_2 \left(\frac{x}{x + 5}\right) = -1$$

Since the base of the log is 2, take 2 and raise it to each whole side:

$$2^{\log_2\left(\frac{x}{x+5}\right)} = 2^{-1}$$

The $\log_2$ is eliminated on the left, leaving the rational expression, and the right is computed to a constant:

$$\frac{x}{x + 5} = 0.5$$

You now have a rational equation, which you will use algebra to solve. Since there is an x in the denominator, find the possible extraneous solution; x cannot equal "-5". Cross-multiply to get:

$0.5x + 2.5 = x$

Subtract 0.5x from both sides to move it to the right and combine like terms:

$0.5x \textbf{- 0.5x} + 2.5 = 1x \textbf{- 0.5x}$

You now have:

2.5 = 0.5x

Divide both sides by 0.5x, and

x = 5

Check this solution by substituting it back into the original equation:

$\log_2 5 = \log_2 (5 + 5) - 1$

$\log_2 5 = \log_2 (10) - 1$

Subtract "$\log_2 (10)$" from both sides:

$\log_2 5 - \log_2 (10) = \log_2 (10) - 1 - \log_2 (10)$

$\log_2 5 - \log_2 (10) = -1$

Consolidate the subtracted logs into a fractional quantity:

$\log_2 \left(\dfrac{5}{10}\right) = -1$

$\log_2 (0.5) = -1$

Convert to exponential form:

$2^{-1} = 0.5$

The "-1" in the exponent moves the 2 to the denominator:

$\dfrac{1}{2} = 0.5$

It checks. Also, the extraneous solution does not conflict with this.

Graphs of Exponentials & Logs

When learning about graphs of exponential and log functions, you should learn, and basically memorize, the 4 parent graphs because all the other graphs are manipulated from these. Pictures of the graphs are not included in this book at this time. Each of the 4 parent functions listed below include valuable information such as the domain, range, and the asymptote the graph approaches:

Exponential graph from:
$f(x) = b^x$, in which $0 < b < 1$
Domain: $(-\infty, \infty)$; Range: $(0, \infty)$
y-intercept: $(0, 1)$
No x-intercept
Asymptote: x-axis
In this case, b is a positive fraction. Please note that similar functions and graphs may be in the form of:
$f(x) = b^{-x}$
which makes a fraction since this causes b to go in the denominator.

Exponential graph from:
$f(x) = b^x$, in which $b > 1$
- $f(x) = e^x$ falls into this category.
Domain: $(-\infty, \infty)$; Range: $(0, \infty)$
y-intercept: $(0, 1)$
No x-intercept
Asymptote: x-axis

Graph of a Log from:
$f(x) = \log_b x$, in which $0 < b < 1$
Domain: $(0, \infty)$; Range: $(-\infty, \infty)$
x-intercept: $(1, 0)$
No y-intercept
Asymptote: y-axis

Graph of a Log from:
$f(x) = \log_b x$, in which $b > 1$
- $f(x) = \ln x$ falls into this category.
Domain: $(0, \infty)$; Range: $(-\infty, \infty)$
x-intercept: $(1, 0)$
No y-intercept
Asymptote: y-axis

Transformations to the Parent Graph/Function

A negative in front of the function, as "-$\log_b$ x" or "-b^x" causes a vertical flip.

A negative in front of the x-quantity, as "$\log_b$ (-x)" or "b^{-x}" causes a horizontal flip.

A coefficient in front of x that is > 1 in "b^x" or the log quantity in "$\log_b$ x" causes a horizontal shrinking effect, making a sharper curve. Whatever that coefficient is, it pushes the line of the graph that many *times* closer to the y-axis. So, if the coefficient is 2, it is pushed twice as close to the y-axis.

A coefficient in front of x that is a positive fraction in "b^x" or the log quantity in "$\log_b$ x" causes a horizontal stretching effect, as it pushes the line of the graph *away* from the y-axis, proportionally.

When b in b^x is > 1, this causes a vertical shrinking effect, pushing the line closer to the x-axis, the higher b is.

A coefficient that is >1 in front of log causes a vertical shrinking effect. A coefficient that is < 1 in front of log causes a vertical stretching effect.

A coefficient that is <1 in front of the log quantity causes a horizontal stretching effect.

The parent graph of an exponential function has a y-intercept of (0, 1) and no x-intercept. For a shift/transformation to an exponential function, you must set x equal to zero then solve for y to find the new y-intercept.

The parent graph of a log function has an x-intercept of (1, 0) and no y-intercept. For a shift/transformation to a log function, you must set "f(x)" (representing y) equal to zero, then solve for x, to find the new x-intercept.

FMMs (Frequently Made Mistakes)

FMM: When Graphing Linear Inequalities, and the inequality is noninclusive, remember to use a dotted line.

FMM: Not isolating the absolute value group first before splitting into 2 inequality statements.

FMM: Thinking that an inequality with no absolute value need not be broken into 2 inequality statements.

FMM: The Wrong Way to Square a Binomial – Squaring only the first and last term (as if you are distributing the exponent 2, but it's not an exponent distribution), and not including the middle term.

FMM: Adding the constant and the coefficient in front of a radical.

FMM: When trying to take the cube root of a number, don't start by dividing by 3 (unless you're finding the cube root of 27... then divide by 3 twice)! I know it may be tempting, but dividing by 3 is treating the number 3 as a factor, not as a root... factors and roots are completely different contexts.

FMM: Assuming that to rationalize a monomial-radical denominator, you only need to multiply it by itself, to eliminate the radical. This is only true for square roots.

FMM: Leaving the sign of the left x-term out of the parentheses. Suppose you are in the middle of doing *factoring by grouping*, with the following factors:
$2x^2 + 2x - 3x - 6$

It is wrong to not keep the sign of each term *inside* the parentheses, like this:
$(2x^2 + 2x) - (3x - 6)$

The correct way is to keep the negative sign associated with 3x inside the parentheses, then to put a plus sign between the sets of parentheses, like this:
$(2x^2 + 2x) + (-3x - 6)$

FMM: Quadradic Inequalities: Not simplifying the inequality first - If one of the simplification steps was multiplying or dividing by zero, the direction of the inequality will change, and substituting the original inequality symbol in for the = sign, to test the test points, will be wrong.

FMMs with logs:
- Improper use of the Product Property
- Improper use of the Quotient Property
- Improper use of the Power Property
- If the variable is in a log quantity (not exponent), be sure to check your solution at the end.

GENERAL TECHNIQUES FOR SUCCESS & IMPROVEMENT

In my first book, *COLLEGE SUCCESS: An Insider's Guide to Higher GRADES, More MONEY, and Better HEALTH* (2010[†]), there is an entire section dedicated to improving grades. One sub-section recommends that students *listen* to the *words* that instructors *say*, and write those notes, because the words spoken often reveal exactly what the instructors want you to know. Students should try to take notes of key words spoken by instructors. This is the essence of *ALGEBRA IN WORDS*.

Keep up with the class. If you miss a class, make sure you study what you missed so that when you return, you aren't clueless. Math and Algebra is hierarchical learning, meaning: the new stuff you learn is dependent on you knowing the previous stuff.

The best ways to study are by
- Doing as many problems and questions as you can, checking your answers, then asking for clarification, and learning from your mistakes. It's ok to make mistakes – it's part of the learning process… but make the mistakes before the test so you don't make as many during the test.
- Getting as much perspective as possible. This includes:
  o Actively participating in class
  o Taking photos of the board. Ask your instructor for permission in the beginning of class or the semester
  o Reading on your own
  o Doing problems and exercises in the book
  o Reviewing and learning from graded tests
  o Using (speaking) the proper terminology
  o Using computer programs
  o Watching videos
  o Writing out the steps *in words*, not just showing the steps. *This book* focuses on that.

When doing problems or taking tests:
- Stay calm
- Ask yourself "What kind of problem is this?" This may lead you to remember what to look for and how to do it.
- Write out the formulas and equations before filling in the given information.
- Map out whatever you know, in an organized way – this might gain you partial credit because it shows you are thinking about it, and it may spark your own thought to solving.

† This book was originally titled *GRADES, MONEY, HEALTH: The Book Every College Student Should Read*. It was re-titled to increase its internet search-abilty.

CLOSING

Thank you for purchasing, reading, and using this guide.

To some, math may seem like a mix of intimidating, complicated, or ambiguous. My goal is to make math concepts accessible and easily understandable. If it works, and these methods reach enough people, I would like to see a generation of students with significantly stronger math scores, especially those from low-income regions. I think it would make the U.S. and the World better.

Please email any feedback to my personal email address:
bullockgr@gmail.com

You can also follow Gregory Bullock on Amazon so you can be notified of future publications,
and on Twitter @GregBullock and @AlgebraInWords.

24303979R00091

Made in the USA
San Bernardino, CA
19 September 2015